Fouad Soliman
Hamed Mira
Islam Alhindawy

Deep Insight for Energy Systems Transformation up to 2060

AF755956

Fouad Soliman
Hamed Mira
Islam Alhindawy

Deep Insight for Energy Systems Transformation up to 2060

LAP LAMBERT Academic Publishing

Imprint
Any brand names and product names mentioned in this book are subject to trademark, brand or patent protection and are trademarks or registered trademarks of their respective holders. The use of brand names, product names, common names, trade names, product descriptions etc. even without a particular marking in this work is in no way to be construed to mean that such names may be regarded as unrestricted in respect of trademark and brand protection legislation and could thus be used by anyone.

Cover image: www.ingimage.com

Publisher:
LAP LAMBERT Academic Publishing
is a trademark of
Dodo Books Indian Ocean Ltd. and OmniScriptum S.R.L publishing group

120 High Road, East Finchley, London, N2 9ED, United Kingdom
Str. Armeneasca 28/1, office 1, Chisinau MD-2012, Republic of Moldova, Europe
Managing Directors: Ieva Konstantinova, Victoria Ursu
info@omniscriptum.com

Printed at: see last page
ISBN: 978-620-8-44462-4

Copyright © Fouad Soliman, Hamed Mira, Islam Alhindawy
Copyright © 2025 Dodo Books Indian Ocean Ltd. and OmniScriptum S.R.L publishing group

Deep Insight for Energy Systems Transformation up to 2060

By

Fouad A. S. Soliman   Hamed I.E.Mira   Islam G. Alhindawy

Nuclear Materials Authoriy,
Ministry of Electricity and Renewable Energy, Cairo, Egypt

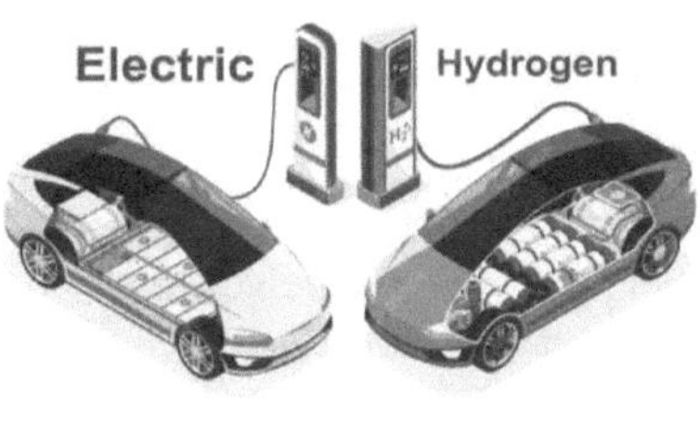

April 2025

About the Authors

Dr. Eng. Fouad A. S. Soliman

Prof. of Electronics and Computers Engineering,
Nuclear Materials Authority, Cairo, Egypt.

Editorial Board Member of:

- **Progress in Photovoltaic, "Research and Applications', John Wiley and Sons, UK, since 1993,**
- **Periodicals of the Association for the Advancement of Modeling and Simulation Techniques, AMSE, Lune, France,**
- **International Journal of Computer Science and Engineering Applications (IJCSEA).**

Member of:

- **American Association for Advancement of Sciences, N.Y., U.S.A,**
- **New York Academy of Sciences, New York, U. S. A.**

Chosen for:

- **Who's Who in the World, A.N. Marquis, N.J., U. S. A.**
- **Outstanding People of the 20th Century, International Biographical Center of Cambridge, England.**

Teaching at Universities

- **Teaching the post-graduate students in the Egyptian Universities.**

Publications and M.Sc. and Ph.D. Supervision

Articles and Supervised Theses

- **Around 200**

Books:

[1]. Fouad A. S. Soliman, **"A Novel Look on the world of Nanotechnology for Today and Future",** Published Book, Lambert Academic Publishing, Omni- Scriptum GmbH and Co. KG, February, 2016, ISBN: 978-3-659-83496-7.

[2]. F. A. S. Soliman, **"Energy and the Future of Civilizations",** Published Book, Lambert Academic Publishing, Omni-Scriptum GmbH and Co.

KG, April 2016.
ISBN: 978-3-659-88129-9.

[3]. **F. A. S. Soliman, "Characterization, Simulation, Applications, Deployment and Economics of Solar Energy",** Lambert Academic Publishing, LAP, Saarbrücken, Germany, May 2016.
ISBN: 978-3-659-89387-2.

[4]. **Fouad A. S. Soliman and Hoda A. Ashry," Role of the Nuclear Technology on Human Daily Life",** Published Book, Lambert Academic Publishing, Omni-Scriptum, GmbH and Co. KG, May 2016.
ISBN: 978-3-659-90461-5.

[5]. **Fouad A. S. Soliman, Safaa M. R. El-ghanam and Ashraf M. Abdel-Maksoud, "Impact of Outer Space Environment on Electronic Devices and Systems",** Published Book, Lambert Academic Publishing, Omni-Scriptum GmbH and Co. KG, July 2016.
ISBN: 978-3-659-93044-7

[6]. **H. A. Ashry, Fouad A. S. Soliman and S. A. Kamh, "Nuclear Technology: Future Generation, Protection and Monitoring", Published Book, Lambert Academic Publishing, Omni-Scriptum GmbH and Co. KG, August 2016.**
ISBN: 978-3-659-93921-1

[7]. **Fouad. A .S. Soliman, "Agriculture in Remote Areas Based on Solar Energy", Published Book,** Lambert Academic Publishing, Omni-Scriptum GmbH & Co. KG, Sept. 2016.
ISBN: 978-3-659-95267-8

[8]. **Fouad A. S. Soliman, "Solar-Wind Hybrid Renewable Energy for Sustainable Agriculture**", Published Book, Lambert Academic Publishing, Omni- Scriptum GmbH and Co. KG, October, 2016, No.: 145917
ISBN: 978-3-659-96384-1

[9]. **Fouad A. S. Soliman, "High Voltage Transmission Lines: Importance, Maintenance and Risks",** Published Book, Lambert Academic Publishing, Omni- Scriptum GmbH and Co. KG, November, 2016, Number:147937,
ISBN: 978-3-330-00309-5.

[10]. **Hoda A. Ashry and Fouad A. S. Soliman, Nuclear Analytical Techniques and Modern Sciences,** Published Book, Lambert Academic Publishing, Omni- Scriptum, GmbH and Co. KG, December, 2016, No.: 149558,
ISBN: 978-3-330-01772-6.

[11]. Fouad A. S. Soliman, **Energy: History, Definitions, Forms, Transformation and Applications,** Published Book, Lambert Academic Publishing, Omni- Scriptum, GmbH and Co. KG, January 2017.
ISBN: 978-3-330-02939-2.

[12]. Fouad A. S. Soliman, **All About Nuclear Materials, Published Book,** Lambert Academic Publishing, Omni- Scriptum GmbH and Co. KG, 2017. Project-ID (150859)
ISBN:978-3-330-03643-7.

[13]. Fouad A. S. Soliman **and Hoda A. Ashry, Focus on the Treasures of The Earth,** Published Book, Lambert Academic Publishing, Omni-Scriptum GmbH and Co. KG, February 2017.
ISBN: 978-3-659-85407-1.

[14]. Fouad A. S. Soliman, **Geothermal Energy Technology,** Published Book, Lambert Academic Publishing, Omni-Scriptum GmbH and Co., KG., May 2017.
ISBN: 978-3-330-31808-3.

[15]. Fouad A. S. Soliman, **Marine Power Technology and Future of Energy,** Published Book, Lambert Academic Publishing, Omni-Scriptum GmbH and Co. KG, June 2017.
ISBN: 978-3-330-32467-1.

[16]. Fouad A. S. Soliman **and Hoda A. Ashry, Atomic Batteries: the Easy Energy for Tomorrow", Published Book,** Lambert Academic Publishing, Omni-Scriptum. GmbH and Co. KG, July 2017.
ISBN:978-3-330-35308-4.

[17]. Fouad A. S. Soliman, **and Hoda A. Ashry, Evolution of Synchrotron Radiation and its Importance",** Published Book, Lambert Academic Publishing, Omni-Scriptum GmbH and Co. KG, August 2017.
ISBN: 978-620-2-01385-7

[18]. Fouad A. S. Soliman, **"Mechatronics: Multidisciplinary Engineering",** Published Book, Lambert Academic Publishing, Omni-Scriptum, GmbH and Co. KG, August 2017.
ISBN: 978-620-0-43740-2.

[19]. Fouad A. S. Solimna **and Hoda A. Ashry, "Gold and Silver Recovery from Electronic Waste", Gold and Silver Recovery from Electronic Waste",** Published Book Lambert Academic Publishing, Omni-Scriptum GmbH and Co. KG, Sep. 2017.
ISBN: 978-620-2-04988-7.

[20]. Fouad A. S. Soliman,**Amira A El-laboudi, and Manal Mahdi, "Harvesting Energy and Future Human Needs",** Published Book,

Lambert Academic Publishing, Omni-Scriptum GmbH and Co. KG, November 2017.
ISBN: 978-620-2-07981-5.

[21]. **Hoda A. Ashry and** Fouad A. S. Soliman, **"World of Neurons",** Published Book, Lambert Academic Publishing, Omni- Scriptum GmbH and Co. KG, January 2018.
ISBN: 978-613-4-97714-2.

[22]. Fouad A. S. Soliman, **"Role of Engineering in Therapy",** Published Book Lambert Academic Publishing, Omni- Scriptum GmbH and Co. KG, April 2018.
ISBN: 978-613-9-58735-3.

[23]. Fouad A. S. Soliman, **"New Trends in Exploring Earth Treasures",** Published Book, Lambert Academic Publishing, Omni- Scriptum GmbH Co. KG, Nov. 2019.
ISBN: 978-620-0-46469-9.

[24]. Fouad A. S. Soliman, **"Energy: Resources, Derivative, Sustainability and Development",** Published Book Lambert Academic Publishing, Omni-Scriptum GmbH and Co. KG, December 2019.

[25]. Fouad A. S. Soliman **and Hamed I. E. Mira, "Nuclear Power: History, Materials, Economics and Future",** Published Book Lambert Academic Publishing, Omni-Scriptum GmbH & Co. KG, January 2020.
ISBN: 978-620-0-46407-1.

[26]. Fouad A. S. Soliman, **"Renewable Energy and the Future of Human Life",** Published Book. Lambert Academic Publishing. Omni-Scriptum GmbH and Co.KG, February 2020.
ISBN: 978-620-0-53632-7.

[27]. Fouad A. S. Soliman, **Safaa M. R. El-ghanam, and Ashraf M. Abdel-maksoud, "Environmental Impact of the Energy Industry",** Published Book Lambert Academic Publishing, Omni-Scriptum GmbH and Co. KG, February 2020.
ISBN: 978-620-0-57165-6.

[28]. Fouad A. S. Soliman, **and Amira Abdel-Magid, "Projections, Developments and Exploitations of Renewable Energy Resources"** Published Book, Lambert Academic Publishing, Omni-Scriptum GmbH and Co. KG, March 2020.
ISBN: 978-620-065158-7.

[29]. Fouad A. A. Soliman, and **Wafaa Abd El-Basit, "Smart Photovoltaic Technologies and the Future of Energy",** Published Book, Lambert Academic Publishing, Omni-Scriptum GmbH and Co. KG, March 2020.

ISBN: 978-620-251267-1.

[30]. **Fouad A. S. Soliman, and Sanaa A. Kamh", Open Source Hardware Technology,** Published Book, Lambert Academic Publishing, Omni-Scriptum GmbH and Co. KG, April 2020.
ISBN: 978-620-2-51639-6.

[31]. **Fouad A. S. Soliman, "Renewable Energy Technologies for Salt Water Desalination",** Published Book, Lambert Academic Publishing, Omni Scriptum GmbH and Co. KG., May 2020.
ISBN: 978-620-2-52159-8.

[32]. **Fouad A. S. Soliman, " New Trends in Renewable Energy for Humanity Benefits",** Published Book, Lambert Academic Publishing, Omni-Scriptum GmbH and Co. KG, May 2020.
ISBN: 978-620-2-51887-1.

[33]. **Fouad A. S. Soliman, and Ashraf M. Abdel-maksoud, "Energy Storage, Transmission and Monitoring",** Published Book, Lambert Academic Publishing, Omni-Scriptum GmbH and Co. KG, May 2020.
ISBN: 978-6213-94971-2

[34]. **Fouad S. S. Soliman, "Climate Effects on PV-Systems and their Maintenance and Recycling",** Published Book, Lambert Academic Publishing, Omni-Scriptum GmbH and Co. KG, June 2020.
ISBN: 978-620-2-56451-9.

[35]. **Fouad A. S. Soliman and Hamed I. E. Mira, "Drones: The Future of Unmanned Aerial Vehicles",** Published Book Lambert Academic Publishing, Omni-Scriptum GmbH and Co. KG, June 2020.
ISBN: 978-620-2-66811-8.

[36]. **Fouad A. S. Soliman, "Airborne Geophysical & Remote Sensing Based on Drone Aircrafts",** Published Book, Lambert Academic Publishing, Omni-Scriptum GmbH and Co. KG, July 2020.
ISBN: 978-620-2-67331-0.

[37]. **Fouad A. S. Soliman and Hamed I. E. Mira, "UXO Environmental Impact, Detection and Demining",** Published Book Lambert Academic Publishing, Omni-Scriptum GmbH and Co. KG, July 2020.
ISBN: 978-620-2-67862-9.

[38]. **Fouad A. S. Soliman, and Safaa M. El-ghanam "The World of Renewable Energy Technologies",** Published Book, Lambert Academic Publishing, Omni-Scriptum GmbH and Co. KG, August 2020.
ISBN: 978-620-2-68432-3.

[39]. **Fouad A. S. Soliman, and Ashraf M. Abedel-maksoud", Technologies of Stand-Alone and Distributed Energy Systems",**

Published Book, Lambert Academic Publishing, Omni-Scriptum GmbH and Co. KG, September 2020.
ISBN: 978-620-0-50455-6.

[40]. **Fouad A. S. Soliman,"A Novel and Efficient Aerial Techniques for UXO Detection",** Published Book, Lambert Academic Publishing, Omni-Scriptum GmbH and Co. KG, September 2020.
ISBN: 978-620-2-79934-8

[41]. **Fouad A. S. Soliman, and Ashraf M. Abedel-maksoud, "Technology and Future of Nano-fluids",** Published Book, Lambert Academic Publishing, Omni-Scriptum GmbH and Co. KG, Sept. 2020.
ISBN: 978-620-2-80132-4.

[42]. **Fouad A. S. Soliman, and Safaa M. El-ghanam,** "New Trends in the Generation, Conversion, Transmission and Storing of Energy", Published Book, Lambert Academic Publishing, Omni-Scriptum GmbH and Co. KG, October 2020.
ISBN: 978-620-2-80878-1.

[43]. **Fouad A. S. Soliman, "Remote Monitoring, Net Metering, Fault Detection and Predictive Maintenance of Electrical Power Systems.** Published Book, Lambert Academic Publishing, Omni-Scriptum GmbH and Co. KG, October 2020.
ISBN: 978-3-330-06474-4.

[44]. **Fouad A. S. Soliman, A. A. Abu Talib and Doaa H. Hanafy, "PV Shockley-Queasier, Maximum Power, Green Houses and Rooftop Stations",** Published Book, Lambert Academic Publishing, Omni-Scriptum GmbH and Co. KG, Oct. 2020.
ISBN: 978-620-2.92085-8.

[45]. **Fouad A. S. Soliman, Wafaa A. Zekri, Soha Abel-Azim, Environ-Mental Impact of Electricity Generation, Transmission and Industry",** Published Book, Lambert Academic Publishing, Omni-Scriptum GmbH and Co. KG, November 2020.
ISBN: 978-620-3-02581-1.

[46]. **Fouad A. S. Soliman, and Safaa R. El-ghanam, "Future Energy DevelopMent",**Published Book, Lambert Academic Publishing, Omni-Scriptum GmbH and Co. KG, November 2020.
ISBN: 978-620-3-041132.

[47]. **Fouad A. S. Soliman and Hamed I. E. Mira, "For More Efficient Solar Energy Applications",** Published Book Lambert Academic Publishing, Omni-Scriptum GmbH and Co. KG, December 2020.
ISBN: 978-620-801002.

[48]. Fouad A. S. Soliman, and **Sanaa A. Kamh, "New Trends in Micro- and Hybrid-Energy Grids"**, Published Book, Lambert Academic Publi**shing**, Omni-Scriptum. GmbH and Co. KG, December 2020. **ISBN: 978-620-2-92022-3.**

[49]. Fouad A. S. Soliman,"**Trends in Renewable Energy Resources Grid ding**", Published Book Lambert Academic Publishing, Omni-Scriptum GmbH and Co. KG, January 2021.
ISBN: 978-620-3-30339-1.

[50]. Fouad A. S. Soliman, and **Wafaa Abdel Basit Zekri, "Gridding of Smart Solar Energy Systems",** Published Book, Lambert Academic Publishing, Omni-Scriptum, GmbH and Co., K.G. March 2021.
ISBN: 978-620-3-46312-5.

[51]. Fouad A. S. Soliman and **Safaa R. El-ghanam, "New Trends in hoto-voltaic System",** Published Book, Lambert Academic Publishing, Omni-Scriptum GmbH and Co., K.G., Dec. 2020.
ISBN: 978-620-3-47075-8.

[52]. Fouad A. S. Soliman, **"Automatic Monitoring of PV-Systems',** Published Book Lambert Academic Publishing, Omni-Scriptum GmbH and Co. KG, Sept. 2021.
ISBN: 978-620-3-58196-6.

[53]. Fouad A. S. Soliman, and **Ashraf M. Abedel-maksoud, "Marine Power: The Future of Renewable Energy,** Published Book, Lambert Academic Publishing, Omni-Scriptum GmbH and Co. KG, November, 2021.
ISBN: 978-620-4-71792-0163.

[54]. Fouad A. S. Soliman, **"Carbon Capture and Sequestration",** Published Book Lambert Academic Publishing, Omni-Scriptum GmbH and Co. KG, November 2021.
ISBN: 978-620-4-72561-1163.

[55]. Fouad A. S. Soliman, **and Hoda A. Ashry,"Role of Electronics and Computer Sciences on Energy Medicine",** Published Book Lambert Academic Publishing, Omni-Scriptum GmbH and Co. KG, Nov. 2021.
ISBN: 978-620-4-727387.

[56]. Fouad A. S. Soliman, **and Nehal Abou-el fotoh Ali, "Future Challenges of Electronics Based on Piezoelectric".** Published Book Lambert Academic Publishing Omni-Scriptum GmbH and Co. KG, December 2021.
ISBN: 978-620-4-70844.

[57]. Fouad A. S. Soliman, **Ayman H. Shanash and Nehal Abou-el fotoh**

Ali, "Sustainale Energy for Human Safety and Luxury", Published Book Lambert Academic Publishing, Omni-Scriptum GmbH and Co. KG, January2022.
ISBN: 978-620-4-73029-1163.

[58]. Fouad A. S. Soliman, and Nehal Abou-el fotoh Ali, "World of Osmo-Tic Phenomenon", Published Book Lambert Academic Publishing, Omni-Scriptum GmbH & Co. KG, January 2021.
ISBN: 978-620-4-73327-2164.

[59]. Fouad A. S. Soliman, Ayman H. Shanash & Nehal Abou-el fotoh Ali, "A Deep Insight into the Future of Energy", Published Book Lambert Academic Publishing, Omni-Scriptum GmbH and Co. KG, Jan. 2022.
ISBN: 978-620-4-73472-9164.

[60]. Fouad A. S. Soliman, Ayman H. Shanash and Nehal Abou-el fotoh Ali, "Transitioning from Fossil Fuels to Renewable Energy", Published Book Lambert Academic. Publishing, Omni-Scriptum GmbH and Co. KG, February 2022.
ISBN: 978-620-4-74114-7164.

[61]. Fouad A. S. Soliman, Ayman H. Shanash and Nehal Abou-el fotoh Ali, "Ocean Thermal Energy Conversion", Published Book Lambert Academic Publishing, Omni-Scriptum GmbH and Co. KG, Feb. 2022.
ISBN: 978-620-4-74278-61.

[62]. Fouad A. S. Soliman, Ayman H. Shanash and Nehal Abou-el fotoh Ali, "The Rapid Movement toward Clean Green World", Published Book Lambert Academic. Publishing, Omni-Scriptum GmbH and Co. KG, February 2022.
ISBN: 9786-204-745 183.

[63]. Fouad A. S. Soliman, Ayman H. Shanash and Nehal Abou-el fotoh Ali, "Renewable Energy Systems Engineering", Published Book Lambert Academic Publishing, Omni-Scriptum GmbH and Co. KG, February 2022.
ISBN: 978-620-4-74716-3.

[64]. Fouad A. S. Soliman, Ayman H. Shanash and Nehal Abou-el fotoh Ali, "From A – To Z–about Renewable Energy", Published Book Lambert Academic Publishing, Omni-Scriptum GmbH and Co. KG, March 2022.
ISBN: 9786-202-053099.

[65]. Fouad A. S. Soliman, Hamed I. E. Mira and Nehal Abou-el fotoh Ali, **"Steps on the Way of Energy Future and Conservation",** Published Book Lambert Academic Publishing, Omni-Scriptum GmbH and Co.

KG, March 2022.
ISBN: 9786-139-448388.

[66]. **Fouad A. S. Soliman**, Nehal Abou-el fotoh Ali & Karima A. Mahmoud, **"Engineering and Comfortable Smart Life"**, Published Book Lambert Academic Publishing, Omni-Scriptum GmbH and Co. KG, March 2022.
ISBN: 978-620-0-24999-91.

[67]. **Fouad A. S. Soliman**, Hoda A. Ashry and Nehal Abou-el fotoh Ali, **"World of Fuel Cells"**, Published Book Lambert Academic Publishing, Omni-Scriptum GmbH and Co. KG, April 2022.
ISBN: 978-620-4-74855-91.

[68]. **Fouad A. S. Soliman**, Nehal Abou-el fotoh Ali and Wafaa A. Zekri, **"Photovoltaic Systems Engineering"**, Published Book Lambert Academic Publishing, Omni-Scriptum GmbH and Co. KG, April 2022.
ISBN: 978-620-4-74893-11.

[69]. **Fouad A. S. Soliman**, Amira A. Abo-talib and Doaa H. Hanafy, **"Role of Electronic Engineering on Automotive and Mechanic Science"**, Published Book Lambert Academic Publishing, Omni-Scriptum, GmbH and Co. KG, May 2022.
ISBN: 978-620-4-75130-61.

[70]. **Fouad A. S. Soliman**, Nihal Abou-alfotoh Ali," **Nano-fiber: The Future of Materials".** Published Book Lambert Academic Publishing, Omni-Scriptum, GmbH and Co. KG., May 2022.
ISBN: 978-620-4-95505-616.

[71]. **Fouad A. S. Soliman**, Sanaa A. Kamh and Doaa H. Hanafy**", The Brilliant Future of Lithium in Energy Storage"**, Published Book Lambert Academic Publishing, Omni-Scriptum, GmbH and Co. KG., May 2022.
ISBN: 978-620-4-98014-0165519.

[72]. **Fouad A. S. Soliman**, and Hamed I. E. Mira, **"Stereo Microscope: the Nano-Imaging Tool of Future"**, Published Book Lambert Academic Publishing, Omni-Scriptum, GmbH and Co. KG, May 2022.
ISBN: 978-620-5489-406.

[73]. **Fouad A. S. Soliman**, Amira A. Abo-talib El-laboudi and Karima A. Mahmoud,**"Future of Energy Hybrid Technologies"**, Published Book Lambert Academic Publishing, Omni-Scriptum, GmbH and Co. KG, May 2022.
ISBN: 978-620-5489-406.

[74]. **Fouad A. S. Soliman**, Wafaa Abdel-basit Zekri and Karima A. Mahmoud, **"The Brilliant Future of Digital Imaging"**, Published Book

Lambert Academic Publishing, Omni-Scriptum, GmbH and Co. KG, Aug. 2022.
ISBN: 978-6205-4956-12.

[75]. Fouad A. S. Soliman, **"Future of Interdisciplinary Sciences",** Publi. Book Lambert Academic Publishing, Omni-Scriptum, GmbH and Co. KG, October 2022.
ISBN: 978-620-5-50245-71.

[76]. Fouad A. S. Soliman, **and Karima A. Mahmoud, "Fewer Losses on Renewable Energy Generation and Applications",** Published Book Lambert Academic Publishing, Omni-Scriptum, GmbH and Co. KG, October 2022.
ISBN: 978-620-4-980669.

[77]. Fouad A. S. Soliman, **Amira Abou-talib El-laboudi and Doaa H. Hassan, "Food Energy",** Published Book, Lambert Academic Publishing, Omni-Scriptum, GmbH and Co. KG, October 2022.
ISBN: 978-620-5-50995-116.

[78]. Fouad A. S. Soliman, **Wafaa Abdel-basit Zekri & Karima A. Mahmoud," The Brilliant World of Graphene",** Published Book Lambert Academic Publishing, Omi-Scriptum, GmbH and Co. KG, October 2022.
ISBN: 978-620-5-51599-016.

[79]. Fouad A. S. Soliman, **Amira A. Abo-talib & Doaa Hanafy," Wind as a Mainstream Renewable Power",,** Published Book Lambert Academic Publishing, Omni-Scriptum, GmbH and Co. KG, Oct. 2022.
ISBN: 978-620-5-52588-316.4

[80]. Fouad A. S. Soliman, **and Karima A. Mahmoud,** "Unmanned Aerial Vehicle Applications and Development toward Few Grams Weight", Published Book Lambert Academic Publishing, Omni-Scriptum, GmbH and Co. KG, October 2022.
ISBN: 978-620-4-980669.

[81]. Fouad A. S. Soliman, **and Karima A. Mahmoud,** "The Benefits of Plastic and its Imminent Dangers to Humanity". Published Book Lambert Academic Publishing, Omni-Scriptum, GmbH and Co. KG, Oct. 2022.
ISBN: 978-620-5622472.

[82]. Fouad A. S. Soliman, **and Karima A. Mahmoud, "Advanced Technologies for Gold Prospection and Mining",** Published Book Lambert Academic Publishing, Omni-Scriptum, GmbH and Co. KG, Feb. 2023.
ISBN: 978-620-6142263.

[83]. Fouad A. S. Soliman, **and Karima A. Mahmoud, "Neuro-linguistic Programing",** Published Book Lambert Academic Publishing, Omni-Scriptum, GmbH and Co. KG, March 2023.
ISBN: 978-620-14432.

[84]. Fouad A. S. Soliman, **and Karima A. Mahmoud, "Future Techniques In Mind Mapping",** Published Book Lambert Academic Publishing, Omni-Scriptum, GmbH, and Co. KG, March 2023.
ISBN: 978-6206-147640.

[85]. Fouad A. S. Soliman, **and Hamid I. E. Mira, "Copper for Bright Future of Renewable Energy",** Published Book Lambert Academic Publishing, Omni-Scriptum, GmbH and Co. KG, March 2023.
ISBN: 978-6206-142263.

[86]. Fouad A. S. Soliman, **Amira A. Abo-talib and Doaa H. Hanafy, Renewable Energy the Power of World by 2050",** Published Book Lambert Academic Publishing, Omni-Scriptum, GmbH and Co. KG, March 2023. April 2023.
ISBN: 978-6206-153573.

[87]. Fouad A. S. Soliman, **and Karima A. Mahmoud, Global Energy Inter-connection and Practice"** Published Book Lambert Academic Publishing Omni-Scriptum, GmbH and Co. KG. April 2023.
ISBN: 978-6206-153573.

[88]. Fouad A. S. Soliman, **Hamid I. E. Mira and Karima A. Mahmoud, "An Insight into World of Wind Energy Technology".** Published Book Lambert Academic Publishing, Omni-Scriptum, GmbH and Co. KG. September 2023.
ISBN: 978-6206-781967.

[89]. Fouad A. S. Soliman, **Wafaa A. Zekri and Karima A. Mahmoud, "Role of Hydrogen in Human Life".** Published Book Lambert Academic Publishing, Omni-Scriptum, GmbH and Co. KG. Sept. 2023.
ISBN: 978-6206-78625-2.

[90]. Fouad A. S. Soliman, **and Karima A. Mahmoud, "Future of Renewable Energy and Storage Techniques".** Published Book Lambert Academic Publishing, Omni-Scriptum, GmbH and Co. KG. September 2023.
ISBN: 978-6206-790570.

[91]. Fouad A. S. Soliman, **Hamid I. E. Mira and Karima A. Mahmoud, "Importance, Poverty, Transmission, Security of the Renewable**

Energy". Published Book Lambert Academic Publishing, Omni-Scriptum, GmbH and Co. KG. September 2023.
ISBN: 978-6206-8433513.

[92]. **Fouad A. S. Soliman, Hamid I. E. Mira and Karima A. Mahmoud, "Toward 100 % Renewable Energy".** Published Book Lambert Academic Publishing, Omni-Scriptum, GmbH and Co. KG. Dec. 2023.
ISBN: 978-620-7-44774-9.

[93]. **Fouad A. S. Soliman, and Karima A. Mahmoud, "Vehicles Operation at Non-polluted Future".** Published Book Lambert Academic Publishing, Omni-Scriptum, GmbH and Co. KG. Dec. 2023.
ISBN: 978-620-7-45399-3.

[94]. **Fouad A. S. Soliman, and Karima A. Mahmoud, "World of Photonics".** Published Book Lambert Academic Publishing, Omni-Scriptum, GmbH and Co. KG. December 2023.
ISBN: 978-620-7-45399-3.

[95]. **Fouad A. S. Soliman, and Karima A. Mahmoud, "Electronics and Computer Sciences for Fair Elections".** Published Book Academic Publishing, Omni-Scriptum, GmbH and Co. KG. Dec. 2023.
ISBN: 978-620-7-474783.

[96]. **Fouad A. S. Soliman, Hamid I. E.Mira and Karima A. Mahmoud, "Phosphates, Phosphoric Acids and Fule Cells".** Published Book Lambert Academic Publishing, Omni-Scriptum, GmbH and Co. KG. Dec. 2023.
ISBN: 978-620-7-484935.

[97]. **Fouad A. S. Soliman, and Karima A. Mahmoud, "Waste Heat Recovery for Power Generation Applications".** Published Book Lambert Academic Publishing, Omni-Scriptum, GmbH and Co. KG. December 2023.
ISBN: 978-620-7-48768-4.

[98]. **Fouad A. S. Soliman, Hamed I. E. Mira, and Karima A. Mahmoud, "Solar Roads",** Published Book. Lambert Academic Publishing, Omni-Scriptum, GmbH and Co. KG. January 2024.
ISBN: 978-620-3-19965-5.

[99]. **Fouad A. S. Soliman, and Karima A. Mahmoud, "Solar Energy Engineering".** Published Book Lambert Academic Publishing, Omni-Scriptum, GmbH and Co. KG, May 2024.
ISBN: 978-620-7-64062-1.

[100]. **Fouad A. S. Soliman, and Karima A. Mahmoud, "New Look to the World of Dark Energy and Materials".**Published Book Lambert

Academic Publishing, Omni-Scriptum, GmbH and Co. KG, June 2024.
ISBN: 978-620-7-64872-6.

[101]. Fouad A. S. Soliman, and Karima A. Mahmoud,"Artificial Intelligence and the Future of Humanity". Published Book Lambert Academic Publi shing, Omni-Scriptum, GmbH and Co. KG, June 2024.
ISBN: 978-620-7-65198-6.

[102]. Fouad A. S. Soliman, and Karima A. Mahmoud, "Technological Road-maps to Net Zero Emissions Target by 2030 and 2050". Published Book Lambert Academic Publishing, Omni - Scriptum, GmbH and Co. KG, June 2024.
ISBN: 978-620-7-809264.

[103]. Fouad A. S. Soliman, Hamed I. E. Mira and Karima A. Mahmoud, "Perovskite for Brilliant Futuer of Solar Cells". Published Book Lambert Academic Publishing, Omni - Scriptum, GmbH and Co. KG, July 2024.
I SBN: 978-620-7-995462.

[104]. Fouad A. S. Soliman and Karima A. Mahmoud, "Role of Neural Networks on the Bright Future of Computer Sciences". Published Book Lambert Academic Publishing, Omni - Scriptum, GmbH and Co. KG, August 2024.
ISBN: 978-620-8-01103-1.

[105]. Fouad A. S. Soliman, Hamed I. E. Mira and Islam G. El-hendawy, "Photovoltaic cell Regenerations and their Developments Research", Published Book Lambert Academic Publishing, Omni - Scriptum, GmbH and Co. KG, September 2024.
ISBN: 978-620-8-11919-5.

[106]. Fouad A. S. Soliman and Karima A. Mahmoud, "World of Hybridization". Published Book Lambert Academic Publishing, Omni –Scriptum, GmbH and Co. KG, October 2024.
ISBN: 978-620-817902.

[107]. Fouad A. S. Soliman and Karima A. Mahmoud, "Laser Technologies and Future Applications". Published Book Lambert Academic Publishing, Omni –Scriptum, GmbH and Co. KG, October 2024.
ISBN: 978-620-8-22485-1.

[108]. Fouad A. S. Soliman, Karima A. Mahmoud and Islam G. Al-hindawy, "Artificial Intelligence for Solar Energy Optimal Generation". Published Book Lambert Academic Publishing, Omni – Scriptum, GmbH and Co. KG, Nov. 2024.

ISBN: 978-3-659--75529-2.

[109]. **Fouad A. S. Soliman, Karima A. Mahmoud,** **Islam Al-hindawy** "All about Ground Solar Station Systems, Rooftop, Floating, and Space". Published Book Lambert Academic Publishing, Omni –Scriptum, GmbH and Co. KG, Nov. 2024.
ISBN: 978-620-4-98142-0.

[110]. **Fouad A. S. Soliman and Karima A. Mahmoud, "Polymer Composites: The Future of High Voltage Insulators"**. Published Book Lambert Academic Publishing, Omni –Scriptum, GmbH and Co. KG, Nov. 2024.
ISBN: 978-3-659-97909-5.

[111]. **Fouad A. S. Soliman and Karima A. Mahmoud,"Optoelectronic Technologies for Human Health, Safety and Luxury"**. Published Book Lambert Academic Publishing, Omni –Scriptum, GmbH and Co. KG, Nov. 2024.
ISBN: 978-3-659-97909-5.

[112]. **Fouad A. S. Soliman and Karima A. Mahmoud,"All about Ground Solar Station Systems, Rooftop, Floating, and Space"**. Published Book Lambert Academic Publishing, Omni –Scriptum, GmbH and Co. KG, Nov. 2024.
ISBN: 978-620-4-98142-0.

[113]. **Fouad A. S. Soliman and Karima A. Mahmoud, "Simple Guide to Energy Generation and Harvesting Technologies for Future"**. Published Book Lambert Academic Publishing, Omni –Scriptum, GmbH and Co. KG, January 2025.
ISBN: 978-620-8-420093.

[114]. **Fouad A. S. Soliman, Hamed I.E.Mira, and Islam G. Alhindawy, "Science and Technologies behind Green Hydrogen Production"** Published Book Lambert Academic Publishing, Omni –Scriptum, GmbH and Co. KG, February 2025.
ISBN: 978-620-8-429089.

[115]. **Fouad A. S. Soliman, and Karima A. Mahmoud, " Breaking Down Solar Farm Costs".** Published Book Lambert Academic Publishing, Omni –Scriptum, GmbH and Co. KG, February 2025.
ISBN: 978-620-8-432867.

[116]. **Islam G. Alhindawy and Fouad A. S. Soliman, "Story of Black Sand: Science, Industry, and Environmental Balance".** Published Book Lambert Academic Publishing, Omni –Scriptum, GmbH and Co. KG,

March 2025.
ISBN: 978-620-8-433994.
[117]. Fouad A. S. Soliman and Karima A. Mahmoud, **"Sustainability of Energy, Engineering and Economics for Human Future".** Published Book. Lambert Academic Publishing, Omni –Scriptum, GmbH and Co. KG, March 2025.
ISBN: 978-620-8-436018.
[118]. Fouad A. S. Soliman **and Karima A. Mahmoud, "Building Integrated Photovoltaic/thermal (BIPV/T) Systems for Future".** Published Book. Lambert Academic Publishing, Omni –Scriptum, GmbH and Co. KG, March 2025.
ISBN: 978-620-8-436629.
[119]. Fouad A. S. Soliman **and Karima A. Mahmoud, "Recent Advancement on the World of Neural Networks".** Published Book. Lambert Academic Publishing, Omni –Scriptum, GmbH and Co. KG, March 2025.
ISBN: 978-620-8-44140-1.
[120]. Fouad A. S. Soliman, **Hamid I. E. Mira and Islam G. Alhindawy, "Achievements in Photovoltaic Technologies within the Next Decades".** Published Book. Lambert Academic Publishing, Omni –Scriptum, GmbH and Co. KG, March 2025.
ISBN: 978-620-44219-4.

Dr. Hamed I. E. Mira

Prof. Geology and Geochemistry
Chairman, Nuclear Materials Authority, Cairo, Egypt.

[1]. **"Nuclear Power: History, Materials, Economics and Future",** Published Book Lambert Academic Publishing,Omni-Scriptum GmbH and Co. KG, January 2020.
ISBN 978-620-0-46407-1.

[2]. **"Drones: The Future of Unmanned Aerial Vehicles",** Published Book Lambert Academic Publishing, Omni-Scriptum GmbH and Co. KG, June 2020.
ISBN: 978-620-2-66811-8.

[3]. **"For More Efficient Solar Energy Applications",** Published Book Lambert Academic Publishing,Omni-Scriptum GmbH and Co. KG, December 2020.
- **ISBN 978-620-801002.**

[4]. **" Solar Roadways: The Future of Renewable Energy",** Published Book Lambert Academic Publishing,Omni-Scriptum GmbH and Co. KG, Jan.2021.
ISBN 978-620-3-19965-9.

[5]. **"Steps on the Way of Energy Future and Conservation",** Published Book Lambert Academic Publishing, Omni-Scriptum GmbH and Co. KG, March 2022.
- **ISBN: 9786-139-448388.**

[6]. **" Stereo Microscope: the Nano-imaging Tool of Future",** Published Book Lambert Academic Publishing, Omni-Scriptum, GmbH and Co. KG, May 2022.
ISBN: 978-620-5489-406.

[7]. **"Copper for Bright Future of Renewable Energy",** Published Book Lambert Academic Publishing, Omni-Scriptum, GmbH and Co. KG, March 2023.
ISBN: 978-6206-142263.

[8]. **"Importance, Poverty, Transmission, Security of the Renewable Energy".** Published Book Lambert Academic Publishing, Omni-Scriptum, GmbH and Co. KG. Sept. 2023.
ISBN: 978-6206-8433513.

[9]. **"An Insight into World of Wind Energy Technology".** Published Book Lambert Academic Publishing, Omni-Scriptum, GmbH and Co. KG. September 2023.
ISBN: 978-6206-781967.

Islam G. Alhindawy
Ph.D. Materials Chemistry

Skills and Expertise:

Material Science, Materials Chemistry, Nanomaterials, Material Characterization, Nanomaterials Synthesis, Sol-Gel, Hydrothermal, Ceramics, Advanced Materials,Smart Materials, Catalysts, Powder Semiconductor, Water Treatment, Adsorption,Photocatalytic Degradation.

Publications:

- I.G. Alhindawy, E.A. Elshehy, M.E. El-Khouly, Y.K. Abdel-Monem, and M.S. Atrees, Fabrication of Mesoporous NaZrP Cation-Exchanger for U(VI) Ions Separation Uranyl Leach Liquors. Colloids and Interfaces. 3(4): p. 61, (2019).

- M.O.A. El-Magied, A.I.L.A.E. Fatah, H. Mashaal, A. Tawfique, I.G. Alhindawy, E.S.A. Manaa, and E.A. Elshehy, Fabrication of Worm-Like Meso-porous Silica Monoliths as an Efficient Sorbent for Thorium Ions from Nitrate Media. Radiochemistry. 64(1): p. 62-73, (2022).

- I.G. Alhindawy, E.A. Elshehy, A.O. Youssef, S.M. Abdelwahab, A.A. Zaher, W.A. El-Said, H.I. Mira, and A.M. Abdelkader, Improving the photo-catalytic performance of cobalt-doped titania nanosheets by induced oxygen vacancies for efficient degradation of organic pollutants. Nano-Structures & Nano-Objects. 31: p. 100888, (2022). https://doi.org/10.1016/j.nanoso.2022.100888

- I.G. Alhindawy, H.I. Mira, A.O. Youssef, S.M. Abdelwahab, A.A. Zaher, W.A. El-Said, E.A. Elshehy, and A.M. Abdelkader, Cobalt doped titania-carbon nanosheets with induced oxygen vacancies for photocatalytic degradation of uranium complexes in radioactive wastes. Nanoscale Advances. 4(24): p. 5330-5342, (2022).https://doi.org/10.1039/D2NA00467D

- A.A. Elzoghby, E.S.A. Haggag, O.E. Roshdy, I.G. Alhindawy, and A.M. Masoud, Kaolinite/thiourea-formaldehyde composite for efficient U(VI) sorption from commercial phosphoric acid. Radiochimica Acta. 111(2): p. 91-103, (2023). https://doi.org/10.1515/ract-2022-0091

- I.G. Alhindawy, D.A. Tolan, E.A. Elshehy, W.A. El-Said, S.M. Abdel-wahab, H.I. Mira, T. Taketsugu, V.P. Utgikar, A.M. El-Nahas, and A.O. Youssef, A novel pH-Dependent sensor for recognition of strontium ions in water: A hierarchically structured mesoporous architectonics. Talanta. 253: p. 124064, (2023).https://doi.org/10.1016/j.talanta.2022.124064.

- I.G. Alhindawy, H. Gamal, A.H. Almuqrin, M.I. Sayyed, and K.A. Mahmoud, Impacts of the calcination temperature on the structural and radiation shielding properties of the NASICON compound synthesized from zircon minerals. Nuclear Engineering and Technology. 55(5): p. 1885-1891, (2023). https://doi.org/10.1016/j.net.2023.02.014.

- K.A. Mahmoud, M.I. Sayyed, A.H. Almuqrin, M.A. Elhelaly, and I.G. Alhindawy, Synthesis of glass powders for radiation shielding applications based on zirconium minerals' leach liquor. Radiation Physics and Chemistry. 207: p. 110867, (2023).https://doi.org/10.1016/j.radphyschem.2023.110867.

- I.G. Alhindawy, M.I. Sayyed, A.H. Almuqrin, and K.A. Mahmoud, Optimizing gamma radiation shielding with cobalt-titania hybrid nanomaterials. Scientific Reports. 13(1): p. 8936, (2023). https://doi.org/10.1038/s41598-023-33864-y.

- D.A. Tolan, A.K. El-Sawaf, I.G. Alhindawy, M.H. Ismael, A.A. Nassar, A.M. El-Nahas, M. Maize, E.A. Elshehy, and M.E. El-Khouly, Effect of bismuth doping on the crystal structure and photocatalytic activity of titanium oxide. RSC Advances. 13(36): p. 25081-25092, (2023). https://doi.org/10.1039/D3RA04034H.

- M.A. Eldoma, S.O. Alaswad, M.A. Mahmoud, I.Y. Qudsieh, M. Hassan, O.Y. Bakather, G.A. Elawadi, A.F.F. Abouatiaa, M.S. Alomar, M.S. Elhassan, I.G. Alhindawy, and Z.M. Ahmed, Enhancing photocatalytic performance of Co-TiO2 and Mo-TiO2-based catalysts through defect engineering and doping: A study on the degradation of organic pollutants under UV light. Journal of Photochemistry and Photobiology A: Chemistry. 446: p. 115164, (2024). https://doi.org/10.1016/ j.jphotochem. 2023.115164.

- I.G. Alhindawy, M.I. Sayyed, D.A. Aloraini, A.H. Almuqrin, M.S. Alomar, G.A. Elawadi, and K.A. Mahmoud, A multi-phase investigation to understand

the function of lanthanum and neodymium in the zirconia ceramics' synthesis, structural, and gamma-ray protective ability. Radiation Physics and Chemistry. 215: p. 111336, (2024).
https://doi.org/10.1016/j.radphyschem.2023.111336.

- I.G. Alhindawy, H. Gamal, A.A. Zaher, M.I. Sayyed, A.H. Almuqrin, D.A. Aloriani, Y.A. Elsheikh, O.Y. Bakather, and K.A. Mahmoud, La/Nd-doped zircon-ium oxide: Impact of zirconia phase transition on gamma-ray shielding properties. Journal of Physics and Chemistry of Solids. 187: p. 111828, (2024).
https://doi.org/ 10.1016/j.jpcs.2023.111828.

- A.K. El-Sawaf, A.A. Nassar, D.A. Tolan, M. Ismael, I. Alhindawy, E. M. El-Desouky, A. El-Nahas, M. Shahien, and M. Maize, A mesoporous Mo and N Co-doped TiO2 nanocomposite with enhanced photocatalytic efficiency. RSC Advances. 14(5): p. 3536-3547, (2024).
https://doi.org/10.1039/D3RA07258D.

- K.A. Mahmoud, M. Binmujlli, M. Marashdeh, M.I. Sayyed, M.J. Aljaafreh, H. Akhdar, and I.G. Alhindawy, Comprehensive analysis of the effects of Mo and Co on the synthesis, structural, and radiation-shielding properties of TiO2 based composites. Progress in Nuclear Energy. 169: p. 105105, (2024). https://doi.org/ 10.1016/j.pnucene.2024.105105.

- M.A. Eldoma, N. Zouli, G.A. Elawadi, M.A. Mahmoud, I.Y. Qudsieh, O.Y. Bakather, M. Hassan, M.S. Alomar, A.F.F. Abouatiaa, S.E.F. Hegazi, Y.A. Elsheikh, K.A. Mahmoud, and I.G. Alhindawy, Utilizing a one-step hydrothermal method for fabrication and evaluation of radiation shielding characteristics in Pb(ZrO3)-doped zirconia: synthesis and characterization. Journal of Materials Science. 59(8): p. 3253-3269, (2024).
https://doi.org/10.1007/s10853-024-09419-5.

- I.G. Alhindawy, M.W. Marashdeh, M.J. Aljaafreh, M. Al-Hmoud, S. Alanazi, and K. Mahmoud, Advanced radiation shielding materials: PbO2-doped zirconia ceramics synthesized through innovative sol-gel method. Nuclear Engineering and Technology. 56(7): p. 2444-2451, (2024).
https://doi.org/10.1016/ j.net.2024.02.001.

- W.M. Youssef, M.M. El-Maadawy, A.M. Masoud, I.G. Alhindawy, and A.E.M. Hussein, Uranium capture from aqueous solution using palm-waste based activated carbon: sorption kinetics and equilibrium. Environmental Monitoring and Assessment. 196(5): p. 428, (2024). https://doi.org/10.1007/s10661-024-12560-y.

- K.A. Mahmoud, M.W. Marashdeh, M. Al-Hmoud, M.J. Aljaafreh, S. Alanazi, and I.G. Alhindawy, The potential of bi2–xZrxO3+x/2@ZrO2 (BZO) as a heavyweight radiation shielding material: Fabrication and characterization using a hydrothermal technique directly from zircon mineral. Progress in Nuclear Energy. 172: p. 105216, (2024). https://doi.org/10.1016/j.pnucene.2024.105216.

- D. Tolan, A. El-Sawaf, A.S.A. Ahmed, A. Nassar, N.M. Mohamed, I.G. Alhindawy, E.A. Elshehy, and V. Utgikar, Enhanced photocatalytic activity of (In–Sr–P) tridoped TiO2/Bi2O3 composite loaded on mesoporous carbon: A facile sol-hydrothermal synthesis approach. Materials Chemistry and Physics. 322: p. 129570, (2024). https://doi.org/10.1016/j.matchemphys.2024.129570.

- I.G. Alhindawy, K.A. Mahmoud, M. Rashad, and M.I. Sayyed, Zirconium tungstate (Zr4W8O32)-doped zirconium dioxide (ZrO2) for gamma ray shielding: an in-depth examination of fabrication, characterizations, and gamma ray attenua-tion properties. Journal of Materials Science. 59(27): p. 12285-12304, (2024). https://doi.org/10.1007/s10853-024-09851-7.

- M.S. Alomar, O.Y. Bakather, N. Zouli, M.A. Eldoma, M.A. Mahmoud, S. E. F. Hegazi, Z.M. Ahmed, D.A. Hassan, A.F.F. Abouatiaa, A. Al Askar, H.I. Mira, K.A. Mahmoud, and I.G. Alhindawy, Solar-driven purification: Removing pharmaceutical contaminants from water using carbon-doped NASICON photo-catalyst. Materials Science in Semiconductor Processing. 185: p. 108901, (2025).

- E.A. Elshehy, M.F. Cheira, I.G. Alhindawy, and A.S.A. Ahmed, 2 – Synthesis methodology for size and shape control of two-dimensional materials, in Functionalization of Two-Dimensional Materials and Their Applications, W.A. El-Sagid and N. Ghany, Editors. Woodhead Publishing. p. 19-57, (2024).

https://doi.org/10.1016/B978-0-323-89955-0.00011-X.

Published Books:

[1]. Fouad A. S. Soliman, Hamed I. E. Mira and Islam G. Alhindawy, "Photo-voltaic cell Regenerations and their Developments Research", Published Book Lambert Academic Publishing, Omni - Scriptum, GmbH and Co. KG, September 2024.
ISBN: 978-620-8-11919-5.

[2]. Fouad A. S. Soliman, Karima A. Mahmoud and Islam Al-hendawy, "Artificial Intelligence for Solar Energy Optimal Generation". Published Book Lambert Academic Publishing, Omni –Scriptum, GmbH and Co. KG, Nov. 2024.
ISBN: 978-3-659-75529-2.

[3]. Fouad Soliman, Karima A. Mahmoud, Islam Al-hindawy "All about Ground Solar Station Systems, Rooftop, Floating, and Space". Published Book Lambert Academic Publishing, Omni –Scriptum, GmbH and Co. KG, Nov. 2024.
ISBN: 978-620-4-98142-0.

Acknowledgments

We are kneeling obsequiousness to **ALLAH** thanking **HIM** for showing us the right way. Without **God** help us efforts would have gone astray. It was through the grace of **God** that we able to acquire this great accomplishment. Thanks also for a person whom I am in love him very much, the **Prophet Mohammed {God's praise and peace upon}**.

Also, we wish to express our deepest gratitude to:

Nuclear Materials Authority, Cairo, Egypt.

- **Prof. Hamid I. E. Mira (President),**
- **Prof. Gehad Saleh (Vice President),**
- **Staff members of the different sectors.**

Women College for Arts, Science and Education, Ain-shams University, Cairo, Egypt.

Staff members of Physics Dept., and Electronics Research Lab.

National Center for Radiation Research and Technology, Cairo, Egypt.

Staff Members of the Radiation Physics Dept.

Abstract

An energy transition (or energy system transformation) is a major structural change to energy supply and consumption in an energy system. Currently, a transition to sustainable energy is underway to limit climate change. Most of the sustainable energy is renewable energy. Therefore, another term for energy transition is renewable energy transition. The current transition aims to reduce green-house gas emissions from energy quickly and sustainably, mostly by phasing-down fossil fuels and changing as many processes as possible to operate on low carbon electricity. A previous energy transition perhaps took place during the Industrial Revolution from 1760 onwards, from wood and other biomass to coal, followed by oil and later natural gas.

After the 1973 oil crisis, the term energy transition was coined by politicians and media. It was popularized by US President Jimmy Carter in his 1977 Address on the Nation on Energy, calling to "look back into history to understand our energy problem. Twice in the last several hundred years, there has been a transition in the way people use energy ... Because we are now running out of gas and oil, we must prepare quickly for a third change to strict conservation and to the renewed use of coal and to permanent renewable energy sources like solar power". The term was later globalised after the 1979 second oil shock, during the **1981** United Nations Conference on New and Renewable Sources of Energy.

From the **1990**s, debates on energy transition have increasingly taken climate change mitigation into account. Parties to the agreement committed "to limit global warming to "well below 2 °C, preferably 1.5 °C compared to pre-industrial levels". This requires a rapid energy transition with a downshift of fossil fuel production to stay within the carbon emissions budget.

Keywords

Energy transition, energy system transformation, major, structural, change, energy, supply, consumption, energy system, currently, transition, sustainable, energy, underway, limit, climate, change, sustainable energy, renewable energy, therefore, another, term, renewable energy transition, current, aim, reduce, green house, gas, emissions, from, energy, quickly, sustainably, mostly, by, phasing-down, fossil, fuels, changing, many, processes, as, possible, operate, low, carbon, electricity, previous, energy, transitions, perhaps, took, places, during, onwards, from, woods, other, biomass, coal, followed, oil, and, later, natural gas, over three-quarters, world's, energy, needs, are, met, burning fossil fuels, but, this, usage, emits, greenhouse gases, production, consumption, responsible, most, human-caused, greenhouse, gas emissions, goals, Paris, agreement, on, climate change, emissions, must be, reduced, as soon as, possible, reach net-zero, mid-century, since, late, renewable energy transitions, also been, driven, rapidly, falling, costs, of, both, solar and and wind power, clean energy, cheaper, than, ever, global, solar, module, prices, fell, percent, less than, batteries, saw, their, best, price, decline, in, seven years, another, benefit, energy transition, potential, reduce, health, environmental, impacts, energy industry, heating, buildings, being, electrified, with, heat, pumps, being, the, most, efficient, technology, by far, improve, flexibility, electrical grids, installation, energy storage, super grids, vital, enable, use, variable, weather-dependent, technologies, however, fossil-fuel, subsidies, slowing, energy transition, broad shift, technologies, behaviour, that, are, needed, replace, one, source, energy, with, another, prime, example, change, from, pre-industrial, system, relying, traditional biomass, wind, water and muscle, power, industrial, system, characterized, pervasive, mechanization, steam power, use, coal, define, glossary, its Sixth, sssessment, report, but, does, define, transition, process, changing, from, one, state, or, condition, to, another, given, period, time, transition, can, occur, in, individuals, firms, cities, regions and nations, can be, based, on, incremental, transformative, change, term, energy, transition, was, coined, politicians, media, was, calling, look back, into, history, understand, our, energy problem, twice, last, several, hundred years, there, has been, way, people, use, energy, because, we, are, now, running, out of, gas and oil, must, prepare, quickly, third change, strict, conservation, renewed use, permanent, renewable energy, sources, like, solar power, term, was, later, globalized, after, the, second oil shock, during, conference, debates, energy transition, have, increasingly, taken, climate, change, mitigation, into, account, parties, agreement, committed, limit, global, warming, and preferably.

Table of Contents

Page No.

Chapter (1)
Energy Transition / Energy System Transformation

1.1. Preface

An energy transition (or energy system transformation) is a major structural change to energy supply and consumption in an energy system. Currently, a transition to sustainable energy is underway to limit climate change. Most of the sustainable energy is renewable energy [1]. Therefore, another term for energy transition is renewable energy transition. The current transition aims to reduce greenhouse gas emissions from energy quickly and sustainably, mostly by phasing-down fossil fuels and changing as many processes as possible to operate on low carbon electricity [2]. A previous energy transition perhaps took place during the Industrial Revolution from 1760 onwards, from wood and other biomass to coal, followed by oil and later natural gas [3, 4].

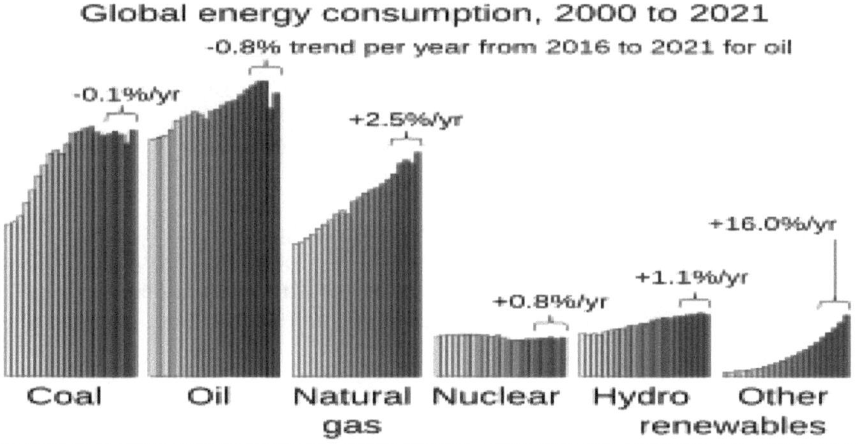

Progress of current energy transition to renewable energy: Fossil fuels such as coal, oil, and natural gas still remain the world's primary energy sources, even as renewables are increasing in use [1].

Over three-quarters of the world's energy needs are met by burning fossil fuels, but this usage emits greenhouse gases [5]. Energy production and consumption are responsible for most human-caused greenhouse gas emissions [6]. To meet the goals of the 2015 Paris Agreement on climate change, emissions must be reduced as soon as possible and reach net-zero by mid-century [7]. Since the late 2010s, the renewable energy transition has also been driven by the rapidly falling cost of both solar and wind power [8]. After 2024, clean energy is cheaper than ever. Global solar module prices fell 35 percent to less than 9 cents/kWh. EV batteries saw their best price decline in seven years [3]. Another benefit of the energy transition is its potential to reduce the health and environmental impacts of the energy industry [9].

Heating of buildings is being electrified, with heat pumps being the most efficient technology by far [10]. To improve the flexibility of electrical grids, the installation of energy storage and super grids are vital to enable the use of variable, weather-dependent technologies [11]. However fossil-fuel subsidies are slowing the energy transition [12, 13].

1.2. Definition

An energy transition is a broad shift in technologies and behaviours that are needed to replace one source of energy with another [14]. A prime example is the change from a pre-industrial system relying on traditional biomass, wind, water and muscle power to an industrial system characterized by pervasive mechanization, steam power and the use of coal.

The IPCC does not define energy transition in the glossary of its Sixth Assessment Report but it does define transition as: "The process of changing from one state or condition to another in a given period of time. Transition can occur in individuals, firms, cities, regions and nations, and can be based on incremental or transformative change "[15].

1.3. Development of the Term

After the 1973 oil crisis, the term energy transition was coined by politicians and media. It was popularized by US President Jimmy Carter in his 1977 Address on the Nation on Energy, calling to "look back into history to understand our energy problem. Twice in the last several hundred years, there has been a transition in the way people use energy ... Because we are now running out of gas and oil, we must prepare quickly for a third change to strict conservation and to the renewed use of coal and to permanent renewable energy sources like solar power " [16]. The term was later globalised after the 1979 second oil shock, during the 1981 United Nations Conference on New and Renewable Sources of Energy [17].

From the 1990s, debates on energy transition have increasingly taken climate change mitigation into account. Parties to the agreement committed "to limit global warming to "well below 2 °C, preferably 1.5 °C compared to pre-industrial levels" [18]. This requires a rapid energy transition with a downshift of fossil fuel production to stay within the carbon emissions budget [19].

1.4. Example of Distributed Generation Use of Renewable Energies

1.4.1. Agricultural Business with Biogas Plant and Photovoltaic Roof

In this context, the term energy transition encompasses a reorientation of energy policy. This could imply a shift from centralized to distributed generation. It also includes attempts to replace overproduction and avoidable energy consumption with energy-saving measures and increased efficiency [20]. The historical transitions from locally supplied wood, water and wind energies to globally supplied fossil and nuclear fuels has induced growth in end-use demand through the rapid expansion of engineering research, education and standardization. The mechanisms for the whole-systems changes include new discipline in Transition Engineering amongst all engineering professions, entrepreneurs, researchers and educators [21].

However it has been argued that the term is a mere slogan and that rather than transitioning, as of 2024, use of all forms of primary energy has increased [22].

1.5. Examples of Past Energy Transitions

Over centuries, energy consumption has evolved from burning wood to fossil fuels (coal, oil, natural gas), and in recent decades to using nuclear, hydroelectric and other renewable energy sources [23].

Historic approaches to past energy transitions are shaped by two main discourses. One argues that humankind experienced several energy transitions in its past, while the other suggests the term "energy additions" as better reflecting the changes in global energy supply in the last three centuries.

The chronologically first discourse was most broadly described by Vaclav Smil [24]. It underlines the change in the energy mix of countries and the global economy. By looking at data in percentages of the primary energy source used in a given context, it paints a picture of the world's energy systems as having changed significantly over time, going from biomass to coal, to oil, and now a mix of mostly coal, oil and natural gas. Until the 1950s, the economic mechanism behind energy systems was local rather than global [25].

The second discourse was most broadly described by Jean-Baptiste Fressoz [26]. It emphasizes that the term "energy transition" was first used by politicians, not historians, to describe a goal to achieve in the future – not as a concept to analyze past trends. When looking at the sheer amount of energy being used by humankind, the picture is one of ever-increasing consumption of all the main energy sources available to humankind [27]. For instance, the increased use of coal in the 19th century did not replace wood consumption, indeed more wood was burned. Another example is the deployment of passenger cars in the 20th century. This evolution triggered an increase in both oil consumption (to drive the car) and coal consumption (to make the steel needed for the car). In other words, according to this approach, humankind never performed a single energy transition in its history but performed several energy additions.

Contemporary energy transitions differ in terms of motivation and objectives, drivers and governance. As development progressed, different national systems became more and more integrated becoming the large, international systems seen today. Historical changes of energy systems have been extensively studied [28]. While historical energy changes were generally protracted affairs, unfolding over many decades, this does not necessarily hold true for the present energy transition, which is unfolding under very different policy and technological conditions [29].

For current energy systems, many lessons can be learned from history [30, 31]. The need for large amounts of firewood in early industrial processes in combination with prohibitive costs for overland transportation led to a scarcity of accessible (e.g. affordable) wood, and eighteenth century glass-works "operated like a forest clearing enterprise" [32]. When Britain had to resort to coal after largely having run out of wood, the resulting fuel crisis triggered a chain of events that two centuries later culminated in the Industrial Revolution [33, 34]. Similarly, increased use of peat and coal were vital elements paving the way for the Dutch Golden Age, roughly spanning the entire 17th century [35]. Another example where resource depletion triggered technological innovation and a shift to new energy sources is 19th century whaling: whale oil eventually became replaced by kerosene and other petroleum-derived products [36]. To speed up the energy transition it is also conceivable that there will be government buyouts or bailouts of coal mining regions [37].

1.6. Drivers for Current Energy Transition

1.6.1. Drivers for Transition to Renewable Energy

With increasing adoption of renewable energy sources, costs have declined, most notably for energy generated by solar panels [38, 39]. (Levelized cost of energy (LCOE) is a measure of the average net present cost of electricity generation for a generating plant over its lifetime.)

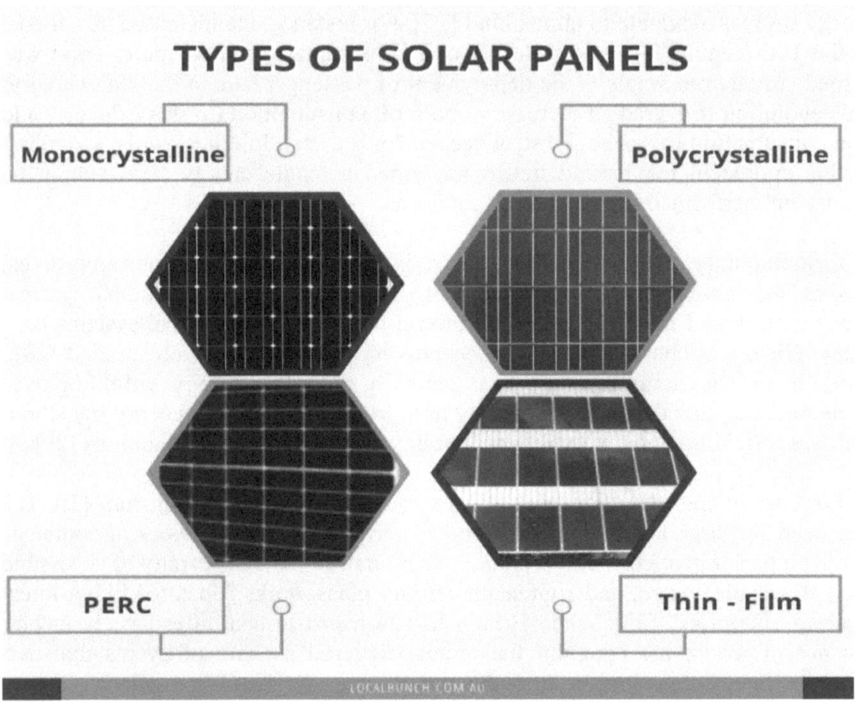

Deaths caused as a result of fossil fuel electricity generation (areas of rectangles in chart) greatly exceed those resulting from production of renewable energy (rectangles barely visible in chart) [40].

1.7. Climate Change Mitigation and Co-benefits

A rapid energy transition to very-low or zero-carbon sources is required to mitigate the effects of climate change [41, 42]. Coal, oil and gas combustion account for 89% of CO2 emissions [43] and still provide 78% of primary energy consumption [44].

Despite the knowledge about the risks of climate change and the increasing number of climate policies adopted since the 1980s, however, energy transitions have not accelerated towards de-carbonization beyond historical trends and remain far off track in achieving climate targets [45].

The deployment of renewable energy can generate co-benefits of climate change mitigation: positive socio-economic effects on employment, industrial development, health and energy access. Depending on the country and the deployment scenario, replacing coal power plants can more than double the number of jobs per average MW capacity [46]. The energy transition could create many green jobs [47], for example in Africa [48, 49]. The costs for retraining workers for the renewable energy industry were found to be trivial for both coal [50] in the U.S. and oil sands in Canada [51]. The latter of which would only demand 2–6% of federal, provincial, and territorial oil and gas subsidies for a single year to be reallocated to provide oil and gas workers with a new career of approximately equivalent pay [51, 52]. In non-electrified rural areas, the deployment of solar mini-grids can significantly improve electricity access [53].

Employment opportunities by the green transition are associated with the use of renewable energy sources or building activity for infrastructure improvements and renovations [54, 55].

1.8. Energy Security

Another important driver is energy security and independence, with increasing importance in Europe and Taiwan [56] because of the 2022 Russian invasion of Ukraine [57]. Unlike Europes **2010s** dependence on Russian gas, even if China stops supplying solar panels those already installed continue generating electricity [58]. Militaries are using and developing electric vehicles, particularly for their stealthiness [59], but not tanks [60]. As of **2023** renewable energy in Taiwan is far too small to help in a blockade [61].

Centralized facilities such as oil refineries [62] and thermal power plants can be put out of action by air attack, whereas although solar can be attacked [63] decentralized power such as solar and wind [64] may be less vulnerable [65, 66]. Solar and batteries reduces risky fuel convoys [67, 68]. However large hydropower plants are vulnerable [69]. Some say that nuclear power plants are unlikely to be military targets [70], but others conclude that civil NPPs in war zones can be weaponized and exploited by the hostile forces not only for impeding energy supplies (and thus shattering the public morale of the adversary) but also for blackmailing and coercing the decision-makers of the attacked state and their international allies with a vision of man-made nuclear disaster [71].

1.9. Economic Development

For many developing economies, for example in the mineral-rich countries of Sub-Saharan Africa, the transition to renewable energies is predicted to become a driver of sustainable economic development. The International Energy Agency (IEA) has identified 37 minerals as critical for clean energy technologies and estimates that by 2050 global demand for these will increase by 235 per cent [72, 73]. Africa has large reserves of many of these so-called "green minerals, such as bauxite, cobalt, copper, chromium, manganese and graphite [74]. The African Union has outlined a policy framework, the Africa Mining Vision, to leverage the continent's mineral reserves in pursuit of sustainable development and socio-economic transformation [75]. Achieving these goals requires mineral-rich African economies to transition from commodity export to manufacture of higher value-added products [76].

1.10. Cost Competitiveness of Renewable Energies

From 2010 to 2019, the competitiveness of wind and solar power substantially increased. Unit costs of solar energy dropped sharply by 85%, wind energy by 55%, and lithium-ion batteries by 85% [77]. This has made wind and solar power the cheapest form for new installations in many regions. Levelized costs for combined onshore wind or solar with storage for a few hours are already lower than for gas peaking power plants [78]. In 2021, the new electricity generating capacity of renewables exceeded 80% of all installed power [79].

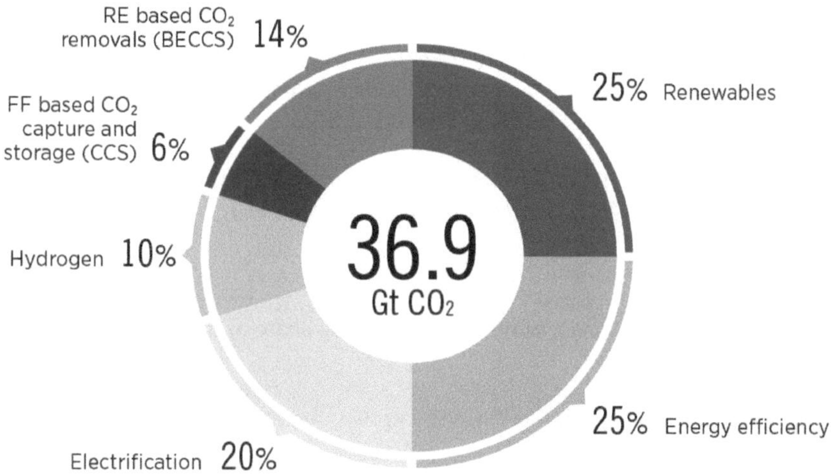

1.11. Key Technologies and Approaches

The emissions reductions necessary to keep global warming below 2 °C will require a system-wide transformation of the way energy is produced, distributed, stored, and consumed [80]. For a society to replace one form of energy with another, multiple technologies and behaviours in the energy system must change [14].

Many climate change mitigation pathways envision three main aspects of a low-carbon energy system:

- The use of low-emission energy sources to produce electricity
- Electrification – that is increased use of electricity instead of directly burning fossil fuels
- Accelerated adoption of energy efficiency measures [81].

1.12. Renewable Energy

Examples of renewable energy options: concentrated solar power with molten salt heat storage in Spain; wind energy in South Africa; the Three Gorges Dam on the Yangtze River in China; biomass energy plant in Scotland.

1.13. Investment

Companies, governments and households have committed increasing amounts to decarbonization, including solar, wind, electric vehicles, charging infrastructure, storage, heating systems, CCS and hydrogen [82-85].

- Bloomberg NEF reported that in 2022, global energy transition investment equaled fossil fuels investment for the first time [86].

1.13.1. Growth in Capacity

- Renewable energy sources provide an increasing share of overall power capacity [87].
- Renewable energy capacity has steadily grown, led by solar photo-voltaic power [88].

In 2023, electricity generation from wind and solar sources was projected to exceed 30% by 2030, as fossil fuels' use continues to decline [89].

Rechargeable battery prices for electric vehicles fell, given economies of scale and new cell chemistries improving energy density [90]. However, general inflationnary pressures, and rising costs of raw materials and components, inhibited price declines in the early 2020s [90].

The most important energy sources in the low carbon energy transition are wind power and solar power. They could reduce net emissions by 4 billion tons CO2 equivalent per year each, half of it with lower net lifetime costs than the reference. Other renewable energy sources include bioenergy, geothermal energy and tidal energy, but they currently have higher net lifetime costs [77].

By 2022, hydroelectricity is the largest source of renewable electricity in the world, providing 16% of the world's total electricity in 2019 [91]. However, because of its heavy dependence on geography and the generally high environmental and social impact of hydroelectric power plants, the growth potential of this technology is limited. Wind and solar power are considered more scalable, but still require vast quantities of land and materials. They have higher potential for growth [92]. These sources have grown nearly exponentially in recent decades thanks to rapidly decreasing costs. In 2019, wind power supplied 5.3% worldwide electricity while solar power supplied 2.6% [91].

While production from most types of hydropower plants can be actively controlled, production from wind and solar power depends on the weather. Electrical grids must be extended and adjusted to avoid wastage. Dammed hydropower is a dispatchable source, while solar and wind are variable renewable energy sources. These sources require dispatchable backup generation or energy storage to provide continuous and reliable electricity. For this reason, storage technologies also play a key role in the renewable energy transition. As of 2020, the largest scale storage technology is pumped storage hydroelectricity, accounting for the great majority of energy storage capacity installed worldwide. Other important forms of energy storage are electric batteries and power to gas.

The " Electricity Grids and Secure Energy Transitions " report by the IEA emphasizes the necessity of increasing grid investments to over $600 billion annually by 2030, up from $300 billion, to accommodate the integration of renewable energy. By 2040, the grid must expand by more than 80 million kilometers to manage renewable sources, which are projected to account for over 80% of the global power capacity increase over the next two decades. Failure to enhance grid infrastructure timely could lead to an additional 58 gigatonnes of CO_2 emissions by 2050, significantly risking a 2°C global temperature rise [93, 94].

1.14. Integration of Variable Renewable Energy Sources

With the integration of renewable energy, local electricity production is becoming more variable. It has been recommended that "coupling sectors, energy storage, smart grids, demand side management, sustainable biofuels, hydrogen electrolysis and derivatives will ultimately be needed to accommodate large shares of renewables in energy systems" [77]. Fluctuations can be smoothened by combining wind and sun power and by extending electricity grids over large areas. This reduces the dependence on local weather conditions.

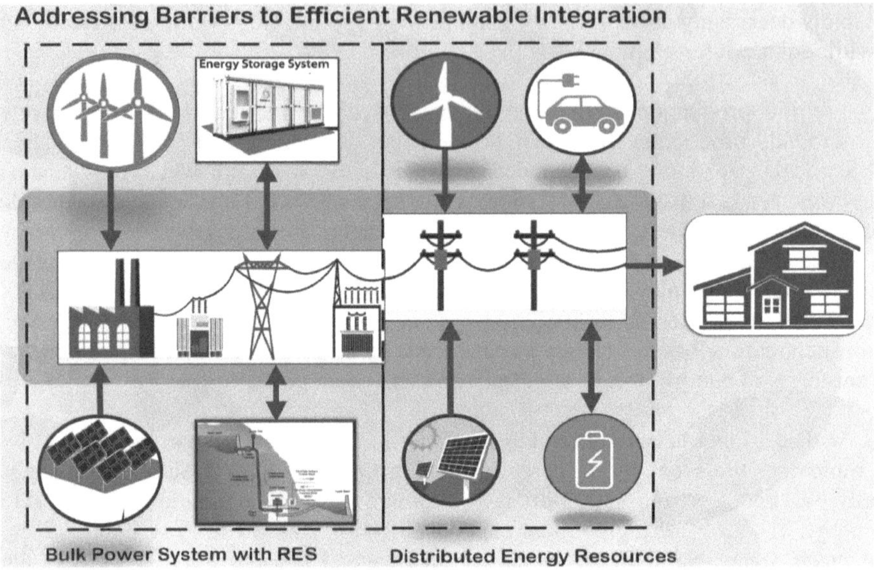

With highly variable prices, electricity storage and grid extension become more competitive. Researchers have found that "costs for accommodating the integration of variable renewable energy sources in electricity systems are expected to be modest until 2030" [77]. Furthermore, "it will be more challenging to supply the entire energy system with renewable energy".

Fast fluctuations increase with a high integration of wind and solar energy. They can be addressed by operating reserves. Large-scale batteries can react within seconds and are increasingly used to keep the electricity grid stable.

1.14.1. 100% Renewable Energy

100% renewable energy is the goal of the use renewable resources for all energy. 100% renewable energy for electricity, heating, cooling and transport is motivated by climate change, pollution and other environmental issues, as well as economic and energy security concerns. Shifting the total global primary energy supply to renewable sources requires a transition of the energy system, since most of today's energy is derived from non-renewable fossil fuels.

Research into this topic is fairly new, with few studies published before 2009, but has gained increasing attention in recent years. The majority of studies show that a global transition to 100% renewable energy across all sectors – power, heat, transport and industry – is feasible and economically viable [95-98]. A cross-sectoral, holistic approach is seen as an important feature of 100% renewable energy systems and is based on the assumption "that the best solutions can be found only if one focuses on the synergies between the sectors" of the energy system such as electricity, heat, transport or industry [99].

The main barriers to the widespread implementation of large-scale renewable energy and low-carbon energy strategies are seen to be primarily social and political rather than technological or economic [100]. The key roadblocks are [101]:

- climate change denial,
- the fossil fuels lobby,
- political inaction, unsustainable energy consumption,
- outdated energy infrastructure, and
- financial constraints.

1.14.2. Nuclear Power

In the 1970s and 1980s, nuclear power gained a large share in some countries. In France and Slovakia more than half of the electrical power is still nuclear. It is a low carbon energy source but comes with risks and increasing costs. Since the late 1990s, deployment has slowed down. Decommissioning increases as many reactors are close to the end of their lifetime or long before because of anti-nuclear sentiments. Germany stopped its last three nuclear power plants by mid April 2023. On the other hand, the China General Nuclear Power Group is aiming for 200 GW by 2035, produced by 150 additional reactors [103].

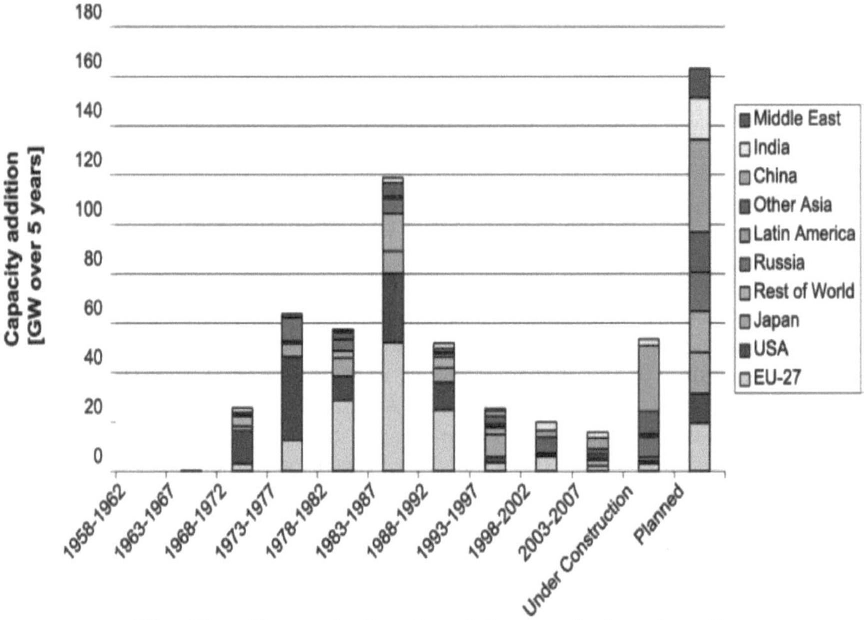

Timeline of commissioned and decommissioned nuclear capacity since the 1958s [102].

1.15. Electrification

With the switch to clean energy sources where power is generated via electricity, end uses of energy such as transportation and heating need to be electrified to run on these clean energy sources. Concurrent with this switch is an expansion of the grid to handle larger amounts of generated electricity to supply to these end uses. Two key areas of electrification are electric vehicles and heat pumps.

It is easier to sustainably produce electricity than it is to sustainably produce liquid fuels. Therefore, adoption of electric vehicles is a way to make transport more sustainable [104]. While electric vehicle technology is relatively mature in road transport, electric shipping and aviation are still early in their development; hence sustainable liquid fuels may have a larger role to play in these sectors [105].

A key sustainable solution to heating is electrification (heat pumps, or the less efficient electric heater). The IEA estimates that heat pumps currently provide only 5% of space and water heating requirements globally, but could provide over 90% [106]. Use of ground source heat pumps not only reduces total annual energy loads associated with heating and cooling, it also flattens the electric demand curve by eliminating the extreme summer peak electric supply requirements [107]. However, heat pumps and resistive heating alone will not be sufficient for the electrification of industrial heat. This because in several processes higher temperatures are required which cannot be achieved with these types of equipment. For example, for the production of ethylene via steam cracking temperatures as high as 900 °C are required. Hence, drastically new processes are required. Nevertheless, power-to-heat is expected to be the first step in the electrification of the chemical industry with an expected large-scale implementation by 2025 [108].

1.16. Economic and Geopolitical Aspects

Countries that managed to reduce their greenhouse gas emissions (working towards a low-carbon economy) while still growing their economy. This is called eco-economic decoupling.

A shift in energy sources has the potential to redefine relations and dependencies between countries, stakeholders and companies. Countries or land owners with resources – fossil or renewable – face massive losses or gains depending on the development of any energy transition. In 2021, energy costs reached 13% of global gross domestic product [109]. Global rivalries have contributed to the driving forces of the economics behind the low carbon energy transition. Technological innovations developed within a country have the potential to become an economic force [110].

1.17. Influences

The energy transition discussion is heavily influenced by contributions from the fossil fuel industries [112]. One way that oil companies are able to continue their work despite growing environmental, social and economic concerns is by lobbying local and national governments.

Historically, the fossil fuel lobby has been highly successful in limiting regulations. From 1988 to 2005, Exxon Mobil, one of the largest oil companies in the world, spent nearly $16 million in anti-climate change lobbying and providing misleading information about climate change to the general public [113]. The fossil fuel industry acquires significant support through the existing banking and investment structure [114]. The concept that the industry should no longer be financially supported has led to the social movement known as divestment. Dives-tment is defined as the removal of investment capital from stocks, bonds or funds in oil, coal and gas companies for both moral and financial reasons [115].

Banks, investing firms, governments, universities, institutions and businesses are all being challenged with this new moral argument against their existing investments in the fossil fuel industry and many; such as Rockefeller Brothers Fund, the University of California, New York City and more; have begun making the shift to more sustainable, eco-friendly investments [116].

In 2024 the International Renewable Energy Agency (IRENA) projected that by 2050, over half of the world's energy will be carried by electricity and over three-quarters of the global energy mix will be from renewables. Although overtaken by both biomass and clean hydrogen, fossil fuels were still projected to supply 12% of energy. The transition is expected to reshape geopolitical power by reducing reliance on long-distance fossil fuel trade and enhancing the importance of regional energy markets [117].

1.18. Social and Environmental Aspects

1.18.1. Impacts

A renewable energy transition can present negative social impacts for some people who rely on the existing energy economy or who are affected by mining for minerals required for the transition. This has led to calls for a just transition, which the IPCC defines as, "A set of principles, processes and practices that aim to ensure that no people, workers, places, sectors, countries or regions are left behind in the transition from a high-carbon to a low carbon economy "[15].

Use of local energy sources may stabilize and stimulate some local economies [118], create opportunities for energy trade between communities, states and regions [119], and increase energy security [120].

Coal mining is economically important in some regions, and a transition to renewables would decrease its viability and could have severe impacts on the communities that rely on this business [121]. Not only do these communities face energy poverty already, but they also face economic collapse when the coal mining businesses move elsewhere or disappear altogether [122]. This broken system perpetuates the poverty and vulnerability that decreases the adaptive capacity of coal mining communities [122]. Potential mitigation could include expanding the program base for vulnerable communities to assist with new training programs, opportunities for economic development and subsidies to assist with the transition [123].

Increasing energy prices resulting from an energy transition may negatively impact developing countries including Vietnam and Indonesia [124].

Increased mining for lithium, cobalt, nickel, copper, and other critical minerals needed for expansion of renewable energy infrastructure has created increased environmental conflict and environmental justice issues for some communities [125, 126].

1.18.2. Labour

A large portion of the global workforce works directly or indirectly for the fossil fuel economy [127]. Moreover, many other industries are currently dependent on unsustainable energy sources (such as the steel industry or cement and concrete industry). Transitioning these workforces during the rapid period of economic change requires considerable forethought and planning. The interna-tional labor movement has advocated for a just transition that addresses these concerns.

Recently [128], an energy crisis is upon the nations of Europe as a result of dependence on Russia's natural gas, which was cut off during the Russia-Ukraine war. This goes to show that humanity is still heavily dependent on fossil fuel energy sources and care should be taken to have a smooth transition, less energy-shortage shocks cripple the very efforts to effectively energize the transition.

1.19. Risks and Barriers

Amongst the key issues to consider in relation to the pace of the global transition to renewables is how well individual electric companies are able to adapt to the changing reality of the power sector. For example, to date, the uptake of renewables by electric utilities has remained slow, hindered by their continued investment in fossil fuel generation capacity [129].

Incomplete regulations on clean energy uptake and concerns about electricity shortages have been identified as key barriers to the energy transition in coal-dependent, fast developing economies such as Vietnam [124].

Researchers found that social sentiments held by U.S. residents have proven to be barriers for energy transitions. The U.S. Department of Energy plans for wind energy to provide 35% of the electrical grid by 2050 are bringing wind energy projects closer to communities. Sentiments for wind energy opposition in local communities include sound annoyance, perceived health effects, and the reduction of scenic landscapes and views.

Anticipated economic aspects are believed to be the most influential variable to perceptions of proposed wind energy developments. Economic barriers to acceptance of renewable energy include local tax increases, increase in electricity rates, decrease in tourism, property value impacts and distributional inequality. However, rural economic development, creation of jobs, investment opportunities, and lower electricity costs stand as possible benefits [130].

1.20. Examples by Country

From 2000 to 2012 coal was the source of energy with the total largest growth. The use of oil and natural gas also had considerable growth, followed by hydropower and renewable energy. Renewable energy grew at a rate faster than any other time in history during this period. The demand for nuclear energy decreased, partly in reaction to a number of high profile accidents (Three Mile Island in 1979, Chernobyl in 1986, and Fukushima in 2011) [131, 132] but also due to the rising cost of nuclear energy [133] which has made it more expensive than all utility scale alternatives [134].

More recently, consumption of coal has declined relative to low carbon energy. Coal dropped from about 29% of the global total primary energy consumption in 2015 to 27% in 2017, and non-hydro renewables were up to about 4% from 2% [135].

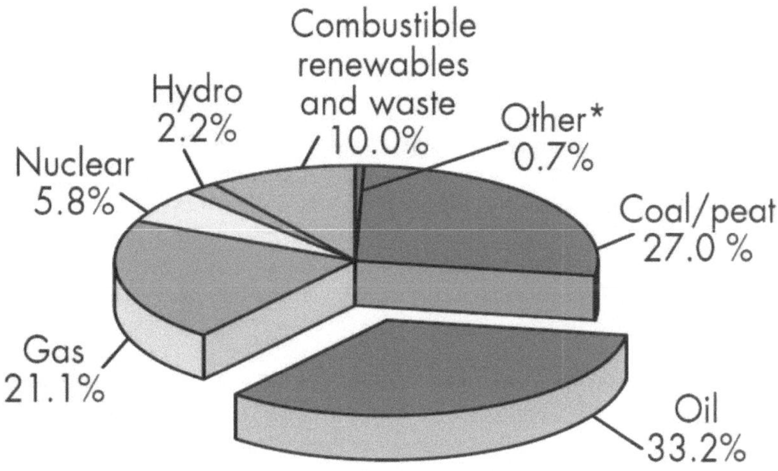

Global energy consumption.

1.20.1 Asia
1.20.1.1. China

Most energy in China comes from coal Development of carbon dioxide emissions in China The 22,500 MW Three Gorges Dam hydroelectric power plant in China, the largest hydroelectric power station in the world.

The People's Republic of China is both the world's largest energy consumer and the largest industrial country. Since the country's industrialization in the 1960s,

China is currently the world's largest emitter of greenhouse gases, and coal in China is a major cause of global warming [136]. China is also the world's largest renewable energy producer, and the largest producer of hydro-electricity, solar power and wind power in the world. The energy policy of China is connected to its industrial policy, where the goals of China's industrial production dictate its energy demand managements [137].

Being a country that depends heavily on foreign petroleum import for both domestic consumption and as raw materials for light industry manufacturing, electrification is a huge component of the Chinese national energy policy.

The Fourteenth Five-Year Plan placed increased emphasis on the green transition as essential to China's pursuit of high-quality and sustainable growth [138].

1.20.1.2. India

India has set renewable energy goals to transition 50% [139] of its total energy consumption into renewable sources in the Paris climate accords. As of 2022 the Central Electricity Authority are well on track of achieving their goals, producing 160 GW electricity from clean sources like solar, wind, hydro power and nuclear power plants, this is 40% of its total capacity. India is ranked third on Ernst and Young's renewable energy country attractive index behind the US and China.

Hydro electric power plants are a major part of India's energy infrastructure since the days of its independence in 1947. Former prime Minister Jawahar Lal Nehru called them the " temples of modern India" and believed them to be key drivers of modernity and industrialism for the nascent republic. Notable examples of hydro power stations include the 2400 MW Tehri hydropower complex, the 1960 MW Koyna hydroelectric project and the 1670 MW Srisailam Dam. Recently, India has given due importance to emerging renewable technologies like solar power plants and wind farms. They house 3 of the world's top 5 solar farms, including world's largest 2255 MW Bhadla Solar Park in and world's second-largest solar park of 2000 MW Pavgada Solar Park and 100 MW Kurnool Ultra mega solar park.

While there has been positive change, air pollution from coal still kills many people [140] and India has to cut down its reliance on traditional coal based power production as it still accounts for around 50% of its energy production. India is also moving towards its goal for electrification of the automotive industry [141] aiming to have at least 30% EV ownership among private vehicles by 2030.

1.20.1.3. Vietnam

Vietnam has led the Southeast Asia in solar and wind uptake, achieving about 20 GW in 2022 from almost zero in 2017 [142]. Thailand has the highest number of EV registrations, with 218,000 in 2022 [143]. The energy transition in Southeast Asia can be summarized as: Challenging, achievable, and interdependent. This implies that while there are obstacles, feasibility largely relies on international support [143].

Public demand for improved local environmental quality and government's aims to promote a green economy are found to be key drivers in Vietnam [142].

Governments ambition to attract international support for green growth initiatives and public demand for a clean environment have been found to be drivers of the energy transition in developing countries, such as Vietnam [144, 145]. Thanks to a relatively more conducive investment environment, Vietnam is poised to a faster energy transition than some other ASEAN members [146].

1.20.2. Europe

1.20.2.1. European Union

The European Green Deal is a set of policy initiatives by the European Commission with the overarching aim of making Europe climate neutral in 2050 [147, 148]. An impact assessed plan will also be presented to increase the EU's greenhouse gas emission reductions target for 2030 to at least 50% and towards 55% compared with 1990 levels. The plan is to review each existing law on its climate merits, and also introduce new legislation on the circular economy, building renovation, biodiversity, farming and innovation [148]. The president of the European Commission, Ursula von der Leyen, stated that the European Green Deal would be Europe's "man on the Moon moment", as the plan would make Europe the first climate-neutral continent [148].

A survey found that digitally advanced companies put more money into energy-saving strategies. In the European Union, 59% of companies that have made investments in both basic and advanced technologies have also invested in energy efficiency measures, compared to only 50% of US firms in the same category. Overall, there is a significant disparity between businesses' digital pro-files and investments in energy efficiency [149].

1.20.2.2. Germany

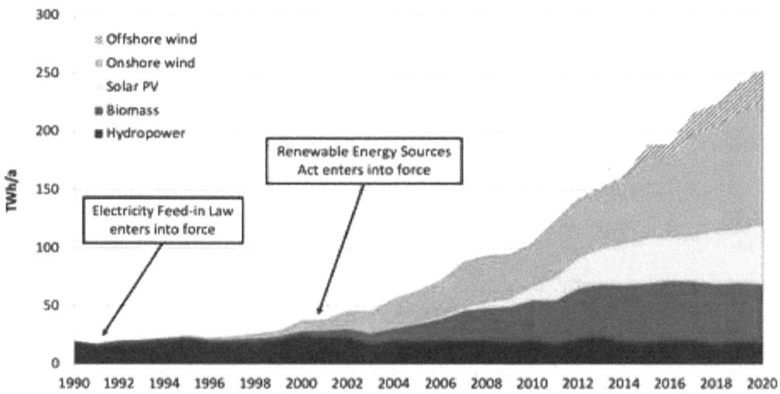

Gross generation of electricity by source in Germany 1990–2020

Germany has played an outsized role in the transition away from fossil fuels and nuclear power to renewables. The energy transition in Germany is known as die Energiewende (literally, "the energy turn") indicating a turn away from old fuels and technologies to new one. The key policy document outlining the Energiewende was published by the German government in September 2010, some six months before the Fukushima nuclear accident; legislative support was passed in September 2010.

The policy has been embraced by the German federal government and has resulted in a huge expansion of renewables, particularly wind power. Germany's share of renewables has increased from around 5% in 1999 to 17% in 2010, reaching close to the OECD average of 18% usage of renewables [150]. In 2022 Germany has a share of 46.2 % and surpassed the OECD average [151]. A large driver for this increase in the shares of renewables energy are decreases in cost of capital. Germany boasts some of the lowest cost of capitals for renewable solar and wind onshore energy worldwide. In 2021 the International Renewable Energy Agency reported capital costs of around 1.1% and 2.4% for solar and wind onshore

[152]. This constitutes a significant decrease from previous numbers in the early 2000s, where capital costs hovered around 5.1% and 4.5% respectively [153]. This decrease in capital costs was influenced by a variety of economic and political drivers. Following the global financial crisis of 2008–2009, Germany eased the refinancing regulations on banks by giving out cheap loans with low interest rates in order to stimulate the economy again [154].

During this period, the industry around renewable energies also started to experience learning effects in manufacturing, project organization as well as financing thanks to rising investment and order volumes. This coupled with various forms of subsidies contributed to a large reduction of the capital cost and the levelized cost of electricity (LCOE) for solar and onshore wind power. As the technologies have matured and become integral parts of the existing sociotechnical systems it is to be expected that in the future, experience effects and general interest rates will be key determinants for the cost-competitiveness of these technologies [153].

Producers have been guaranteed a fixed feed-in tariff for 20 years, guaranteeing a fixed income. Energy co-operatives have been created, and efforts were made to decentralize control and profits. The large energy companies have a disproportionately small share of the renewables market. Nuclear power stations were closed, and the existing nine stations will close earlier than necessary, in 2022.

The reduction of reliance on nuclear stations has had the consequence of increased reliance on fossil fuels. One factor that has inhibited efficient employment of new renewable energy has been the lack of an accompanying investment in power infrastructure to bring the power to market. It is believed 8300 km of power lines must be built or upgraded [150].

Different Länder have varying attitudes to the construction of new power lines. Industry has had their rates frozen and so the increased costs of the Energiewende have been passed on to consumers, who have had rising electricity bills. Germans in 2013 had some of the highest electricity costs in Europe [155]. Nonetheless, for the first time in more than ten years, electricity prices for household customers fell at the beginning of 2015 [156].

1.20.2.3. Switzerland

Due to the high share of hydroelectricity (59.6%) and nuclear power (31.7%) in electricity production, Switzerland's per capita energy-related CO2 emissions are 28% lower than the European Union average and roughly equal to those of France. On 21 May 2017, Swiss voters accepted the new Energy Act establishing the 'energy strategy 2050'. The aims of the energy strategy 2050 are: to reduce energy consumption; to increase energy efficiency; and to promote renew-able energies (such as water, solar, wind and geothermal power as well as biomass fuels) [157]. The Energy Act of 2006 forbids the construction of new nuclear power plants in Switzerland [157].

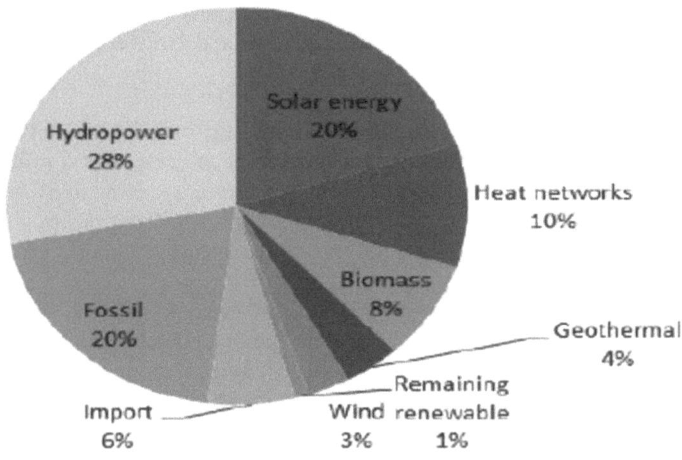

Switzerland electricity generation by source - percentage share

1.20.2.4. United Kingdom

In order to meet the net zero target energy networks must be strengthened [160]. Electricity is only a part of energy in the United Kingdom, so natural gas used for industrial and residential heat [161] and petroleum used for transport in the United Kingdom must also be replaced [162] by either electricity or another form of low-carbon energy, such as sustainable bioenergy crops [163] or green hydrogen [164].

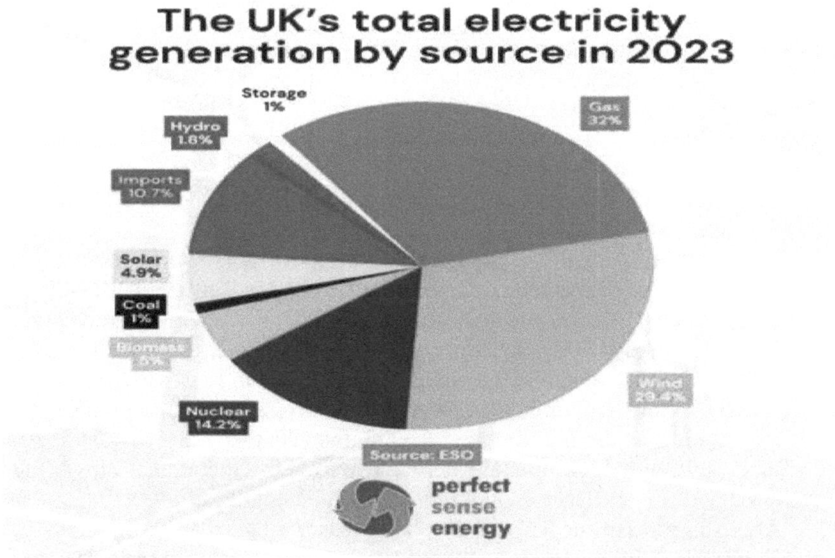

United Kingdom electricity generation by source - percentage share

Although the need for the energy transition is not disputed by any major political party, in 2020 there is debate about how much of the funding to try and escape the COVID-19 recession should be spent on the transition, and how many jobs could be created, for example in improving energy efficiency in British housing [165]. Some believe that due to post-covid government debt that funding for the transition will be insufficient [166]. Brexit may significantly affect the energy transition, but this is unclear as of 2020 [166]. The government is urging UK business to sponsor the climate change conference in 2021, possibly including energy companies but only if they have a credible short-term plan for the energy transition [166].

1.21. References

[1]. Andrew, Robbie. "Figures from the Global Carbon Budget 2021". Retrieved 2022-05-22.
[2]. Tian, Jinfang; Yu, Longguang; Xue, Rui; Zhuang, Shan; Shan, Yuli (2022).

"Global low-carbon energy transition in the post-COVID-19 era". Applied Energy. 307: 118205.
[3]. Davidsson, Simon (2015). "Global Energy Transitions".
[4]. Smil, Vaclav. "Energy Transitions". Archived on 2023-03-09. Retrieved 2022-06-07.
[5]. "Fossil Energy". Center on Global Energy Policy at Columbia University SIPA | CGEP. Retrieved 2024-04-16.
[6]. "Greenhouse Gas Emissions from Energy Data Explorer – Data Tools". IEA. Retrieved 2024-04-16.
[7]. "The Paris Agreement". United Nations Framework Convention on Climate Change. Archived from the original on 2021-03-19. Retrieved 2021-09-18.
[8]. "Plunging cost of wind and solar marks turning point in energy transition: IRENA". Reuters. 2020. Archived on 2020-08-10. Retrieved 2020-06-02.
[9]. "Life Cycle Assessment of Electricity Generation Options" (PDF). United Nations Economic Commission for Europe. 2021. pp. 49–55.
[10]. Kök, A. Gürhan; Shang, Kevin; Yücel, Safak (2020-01-23). "Investments in Renewable and Conventional Energy: The Role of Operational Flexibility". Manufacturing & Service Operations Management. 22 (5): 925–941.
[11]. "Abolishing fossil fuel subsidies: a brain teaser rather than a no-brainer". Reforming fossil fuel subsidies is a complex task for politicians. All in all, our study shows that abolishing fossil fuel subsidies is a no-brainer only for a limited number of subsidies. Abolishing inventoried fossil subsidies does not appear to help the energy transition in all cases. It is important to assess policies from the perspective of adequate pricing of climate damage and other externalities.
[12]. Tripathi, Bhasker. "How fossil fuel subsidies are hurting the energy transition | Context". www.context.news. Retrieved 2024-04-16.
[13]. Jaccard, Mark (2020). "Chapter 11 – "Renewables Have Won"". The Citizen's Guide to Climate Success: Overcoming Myths that Hinder Progress. Cambridge University Press. ISBN 978-1-108-47937-0. Archived on 2021-09-12.
[14]. IPCC, 2022: Annex I: Glossary. In IPCC, 2022: Climate Change 2022: Mitigation of Climate Change. Contribution of Working Group III to the Sixth Assessment Report of the Intergovernmental Panel on Climate Change [P.R. Shukla, et al. J.]. Cambridge University Press, Cambridge, UK and New York, NY, USA.
[15]. Carter, Jimmy. "Address to the Nation on Energy". The American Presidency Project. UC Santa Barbara. Retrieved 2022-06-19.
[16]. Basosi, Duccio (2020-03-20). "Lost in transition. The world's energy past,

present and future at the 1981 United Nations Conference on New and Renewable Sources of Energy". Journal of Energy History. 4. Retrieved 2022-06-19.
[17]. "The Paris Agreement". UNFCCC. Retrieved 2021-01-02.
[18]. Rogelj, Joeri; et al. (July 2019). "Estimating and tracking the remaining carbon budget for stringent climate targets". Nature. 571 (7765): 335–342. ISSN 1476-4687.
[19]. Louis Boisgibault, Fahad Al Kabbani (2020): Energy Transition in Metropolises, Rural Areas and Deserts. Wiley - ISTE. (Energy series) ISBN 9781786304995.
[20]. Krumdieck, Susan (2020). Transition Engineering, Building a Sustainable Future. CRC Press. ISBN 978-0-367-34126-8.
[21]. Fressoz, Jean-Baptiste (2024-10-03). More and More and More.
[22]. "Renewable energy explained". eia.gov. U.S. Energy Information Administration. April 2022. Archived on 2023-03-13. Data source: U.S. Energy Information Administration, Monthly Energy Review, Appendix D.1, and Tables 1.3 an 10.1, April 2022, preliminary data for 2021. / Wood includes wood and wood waste; Renewables includes biofuels, geothermal, solar and wind.
[23]. Smil, Vaclav (2010). Energy Transitions. History, Requirements, Prospects. Praeger. ISBN 978-0-313-38177-5.
[24]. Häfelse, W; Sassin, W (1977). "The global energy system". Annual Review of Energy. 2: 1–30.
[25]. Fressoz, Jean-Ba (2014). "Pour une histoire désorientée de l'énergie". HAL Open Science. Retrieved 2022-03-12.
[26]. "Figure 1: World Energy Consumption by Source, based on Vaclav Smil".
[27]. Höök, Mikael; Li, Junchen; Johansson, Kersti; Snowden, Simon (2011). "Growth Rates of Global Energy Systems and Future Outlooks". Natural Resources Research. 21 (1): 23–41.
[28]. Sovacool, Benjamin K. (2016-03-01). "How long will it take? Conceptualizing the temporal dynamics of energy transitions". Energy Research & Social Science. 13: 202–215. ISSN 2214-6296.
[29]. Podobnik, B. (1999). "Toward a sustainable energy regime: a long-wave interpretation of global energy shifts". Technological Forecasting and Social Change. 62 (3): 155–172.
[30]. Rühl, C.; Appleby, P.; Fennema, F.; Naumov, A.; Schaffer, M. (2012). "Economic development and the demand for energy: a historical perspective on the next 20 years". Energy Policy. 50: 109–116.
[31[. Debeir, J.C.; Deléage, J.P.; Hémery, D. (1991). In the Servitude of Power:

Energy and Civilisation Through the Ages. London: Zed Books. ISBN 9780862329426.
[32]. Nef, J.U (1977). "Early energy crisis and its consequences". Scientific American. 237 (5): 140151.
[33]. Fouquet, R.; Pearson, P.J.G. (1998). "A thousand years of energy use in the United Kingdom". The Energy Journal. 19 (4): 1–41.
[34]. Unger, R.W. (1984). "Energy sources for the dutch golden age: peat, wind, and coal". Research in Economic History. 9: 221–256.
[35]. Bardi, U. (2007). "Energy prices and resource depletion: lessons from the case of whaling in the nineteenth century". Energy Sources, Part B: Economics, Planning, and Policy. 2 (3): 297–304. Archived on 2021-06-24. Retrieved 2020-03-23.
[36]. "The Great Carbon Arbitrage: Going short on coal and long on renewables | Stanford Institute for Economic Policy Research (SIEPR)". siepr.stanford.edu. Retrieved 2023-03-29.
[37]. Chrobak, Ula; Chodosh, Sara (2021-01-28). "Solar power got cheap. So why aren't we using it more?". Popular Science. Archived on 2021-01-29.
[38]. "Lazard LCOE Levelized Cost Of Energy+". Lazard. June 2024. p. 16. Archived on 2024-08-28.
[39]. Ritchie, Hannah; Roser, Max (2021). "What are the safest and cleanest sources of energy?". Our World in Data. Archived on 2024-01-15. Data sources: Markandya & Wilkinson (2007); UNSCEAR (2008; 2018); Sovacool et al. (2016); IPCC AR5 (2014); Pehl et al. (2017); Ember Energy (2021).
[40]. M. Pathak, et al. (2022): Technical Summary. In: Cambridge University Press, Cambridge, UK and New York, NY, USA. doi: 10.1017/9781009157926.002.
[41]. UN Energy (2021) Theme Report on Energy Transition towards the Achievement of SDG 7 and Net-Zero Emissions
[42]. Olivier, J.G.J.; Peters, J.A.H.W. (2020). "Trends in global CO2 and total greenhouse gas emissions (2020)". The Hague: PBL Netherlands Environmental Assessment Agency.
[43]. "Statistical Review of World Energy 2021". BP. Retrieved 2022-05-29.
[44]. Suzuki, Masahiro; Jewell, Jessica; Cherp, Aleh (2023-11-09). "Have climate policies accelerated energy transitions? Historical evolution of electricity mix in the G7 and the EU compared to net-zero targets". Energy Research & Social Science. 106:103281.
[45]. IASS/Green ID (2019). "Future skills and job creation through renewable energy in Vietnam. Assessing the co-benefits of decarbonising the power

sector".
[46]. "Realizing the Green Jobs Promise". RMI. Retrieved 2024-04-18.
[47]. "Renewables set to energize Africa's growth". UNEP. 2018-01-16. Retrieved 2024-04-18.
[48]. "Renewable energy offers Africa's best opportunity to achieve the Sustainable Development Goals, experts say".
[49]. Louie, Edward P.; Pearce, Joshua M. (2016-06-01). "Retraining investment for U.S. transition from coal to solar photovoltaic employment". Energy Economics. 57: 295–302. ISSN 0140-9883.
[50]. Meyer, Theresa K.; Hunsberger, Carol; Pearce, Joshua M. (202). "Retraining investment for Alberta's oil and gas workers for green jobs in the solar industry". Carbon Neutrality. 2 (1): 28. ISSN 2731-3948.
[51]. "How to ensure Alberta's oil and gas workers have jobs during the energy transition - Alberta News". 2023-11-01. Retrieved 2023-12-20.
[52]. IASS/TERI. "Secure and reliable electricity access with renewable energy mini-grids in rural India. Assessing the co-benefits of decarbonising the power sector" .
[53]. "Regional Cohesion in Europe 2021-2022". EIB.org. Retrieved 2022-08-09.
[54]. "Press corner". European Commission - European Commission. Retrieved 2022-08-16.
[55]. Taplin, Nathaniel. "For Taiwan, as for Ukraine, Energy Security Is Existential". WSJ. Retrieved 2024-04-18.
[56]. "EURACTIV Press Release Site". EURACTIV PR. Retrieved 2022-05-22.
[57]. "Europe is importing a solar boom. Good news for (nearly) everyone". The Economist. ISSN 0013-0613. Retrieved 2024-04-18.
[48]. Toll, Micah (2022-05-12). "This quiet electric motorcycle is helping Ukrainian snipers fight back against the Russians". Electrek. Retrieved 2024-04-18.
[59]. Farley, Robert (2023-11-11). "Trump's False Claim That U.S. Military Moving to Electric Tanks". FactCheck.org. Retrieved 2024-04-18.
[60]. https://www.cfe-dmha.org/LinkClick.aspx?fileticket=sJ7hhDPJFl8%3D=0
[61]. Global Taiwan Institute (2024-02-07). "Geopolitics and Energy Security in Taiwan: A Refined Analysis". Global Taiwan Inst. Retrieved 2024-04-18.
[62]. "Ukrenergo: Russia targeted solar power plant in Ukraine's rear for the first time". Yahoo News. 2024-04-04. Retrieved 2024-04-18.
[63]. "Taiwan war game exposes vulnerability of energy grid to a China attack". www.ft.com. Retrieved 2024-04-18.
[64]. "Solar: the military's secret weapon" (PDF).
[65]. "Ukraine and Kyrgyzstan. How solar power can help communities in crisis -

CANEECCA: Восточная Европа, Кавказ и Центральная Азия". caneecca.org. 2022-10-23. Retrieved 2024-04-18.
[66]. "Solar-Powered Defense: How Renewable Energy is Shaping Modern Military Operation". Foresight Learning. 2023. Retrieved 2024-04-18.
[67]. Roza, David (2023-10-31). "Why the Military May Need Microgrids to Win A Near-Peer Fight". Air & Space Forces Magazine. Retrieved 2024-04-18.
[68]. "Ukraine says Russia's attacks are threatening energy security". Voice of America. 2024-03-29. Retrieved 2024-04-18.
[69]. "Taiwan's Election Has Big Ramifications for Energy Security". thediplomat.com. Retrieved 2024-04-18.
[70]. Przybylak, Joanna (2023-11-27). "Nuclear power plants in war zones: Lessons learned from the war in Ukraine". Security and Defence Quarterly.
[71]. "Critical Minerals – Topics". IEA. Retrieved 2024-06-10.
[72]. "Global Critical Minerals Outlook 2024 – Analysis". IEA. 2024-05-17. Retrieved 2024-06-13.
[73]. "Yesterday mineral supplier, tomorrow battery producer - The Nordic Africa Institute". nai.uu.se. Retrieved 2024-06-10.
[74]. "AMV – Africa Mining Vision | African Union". au.int. Retrieved 2024-06-10.
[75]. "Yesterday mineral supplier, tomorrow battery producer - The Nordic Africa Institute". nai.uu.se. Retrieved 2024-06-10.
[76]. IPCC, 2022: Summary for Policymakers. In: Climate Change 2022: Mitigation of Climate Change. Contribution of Working Group III to the Sixth Assessment Report of the Intergovernmental Panel on Climate Change [P.R. Shukla, et al.]. Cambridge University Press, Cambridge, UK and New York, NY, USA. doi: 10.1017/9781009157926.001
[77]. "Levelized Cost of Energy+". www.lazard.com. Retrieved 2024-08-06.
[78]. IRENA (2022), Renewable Energy Statistics 2022, The International Renewable Energy Agency, Abu Dhabi
[79]. United Nations Environment Programme (2019). Emissions Gap Report 2019. UNEP, Nairobi.
[80]. IPCC (2014). Edenhofer; et al. (eds.). Climate Change 2014: Mitigation of Climate Change: Working Group III contribution to the Fifth Assessment Report of the Intergovernmental Panel on Climate Change. Cambridge University Press. ISBN 978-1-107-05821-7. Archived on 2017-01-26.
[81]. "Energy Transition Investment Hit $500 Billion in 2020 – For First Time". BloombergNEF. (Bloomberg New Energy Finance). 2021-01-19. Archived on 2021-01-19.
[82]. Catsaros, Oktavia (2023-01-26). "Global Low-Carbon Energy Technology

Investment Surges Past $1 Trillion for the First Time". Bloomberg NEF (New Energy Finance). Figure 1. Archived on 2023-05-22. Defying supply chain disruptions and macroeconomic headwinds, 2022 energy transition investment jumped 31% to draw level with fossil fuels
[83]. "Global Clean Energy Investment Jumps 17%, Hits $1.8 Trillion in 2023, According to BloombergNEF Report". BNEF.com. Bloomberg NEF. 2024-01-30. Archived on 2024-06-28. Start years differ by sector but all sectors are present from 2020 onwards.
[84]. 2024 data: "Energy Transition Investment Trends 2025 / Abridged report". BloombergNEF. 2025-01-30. p. 9. Archived on 2025-02-02.
[85]. "Energy Transition Investment Now On Par with Fossil Fuel". Bloomberg NEF (New Energy Finance). 2023-02-10. Archived on 2023-03-27.
[86]. "Share of cumulative power capacity by technology, 2010-2027". IEA.org. International Energy Agency (IEA). 2022-12-05. Archived on 2023-02-04. Source states "Fossil fuel capacity from IEA (2022), World Energy Outlook 2022. IEA. Licence: CC BY 4.0."
[87]. Source for data beginning in 2017: "Renewable Energy Market Update Outlook for 2023 and 2024". IEA.org. International Energy Agency (IEA). June 2023. p. 19. Archived on 2023-07-11. IEA. CC BY 4.0. • Source for data through 2016: "Renewable Energy Market Update / Outlook for 2021 and 2022" . IEA.org. International Energy Agency. May 2021. p. 8. Archived on 2023-03-25. IEA. Licence: CC BY 4.0
[88]. Bond, Kingsmill; et al. (2023). "Report / 2023 / X-Change: Electricity / On track for disruption". Rocky Mountain Institute. Archived on 2023-07-13.
[89]. "Race to Net Zero: The Pressures of the Battery Boom in Five Charts". Bloomberg NEF. 2022-07-21. Archived on 2023-09-07.
[90]. "Electricity production". IEA. Retrieved 2022-06-20.
[91]. "Solar Energy Potential". Energy.gov. Archived on 2020-05-23. Retrieved 2020-04-22.
[92]. "Electricity Grids and Secure Energy Transitions – Analysis". IEA. 2023-10-17. Retrieved 2024-04-15.
[93]. IEA (2023), Electricity Grids and Secure Energy Transitions, IEA, Paris https://www.iea.org/reports/electricity-grids-and-secure-energy-transitions, Licence: CC BY 4.0
[94]. Bogdanov, D.; Gulagi, A.; Fasihi, M.; Breyer, C. (2021). "Full energy sector transition toward 100% renewable energy supply: Integrating power, heat, transport and industry sectors including desalination". Applied Energy. 283: 116273. ISSN 0306-2619.
[95]. Teske, Sven, ed. (2019). Achieving the Paris Climate Agreement Goals.

ISBN 978-3-030-05842-5.
[96]. "Cheap, safe 100% renewable energy possible before 2050, Finnish study". Yle Uutiset. 2019-04-12. Retrieved 2021-06-18.
[97]. Gulagi, Ashish; et al. (2021-07-01). "Transition pathway towards 100% renewable energy across the sectors of power, heat, transport, and desalination for the Philippines". Renewable and Sustainable Energy Reviews. 144: 110934. ISSN 1364-0321.
[98]. Hansen, Kenneth; et al. (2019). "Status and perspectives on 100% renewable energy systems". Energy. 175: 471–480. The great majority of all public-cations highlights the technical feasibility and economic viability of 100% RE systems.
[99]. Koumoundouros, Tessa (2019-12-27). "Stanford Researchers Have an Exciting Plan to Tackle The Climate Emergency Worldwide". ScienceAlert. Retrieved 2020-01-05.
[100]. Wiseman, John; et al. (April 2013). "Post Carbon Pathways". University of Melbourne.
[101]. "Operational & Long-Term Shutdown Reactors". IAEA. 2013-04-13. Retrieved 2013-04-14.
[102]. Murtaugh, Dan; Krystal, Chia (2021-11-02). "China's Climate Goals Hinge on a $440 Billion Nuclear Buildout". Bloomberg. Retrieved 2022-07-31.
[103]. Bogdanov, D.; Farfan, J.; Sadovskaia, K.; Aghahosseini, A.; et al. (2019). "Radical transformation pathway towards sustainable electricity via evolutionary steps". Nature Communications. 10 (1): 1077.
[104]. "World Energy Outlook 2020 – Analysis". IEA. 2020-10-13. Retrieved 2024-08-06.
[105]. Abergel, Thibaut (2020). "Heat Pumps". IEA. Archived on 2021-03-03. Retrieved 2021-04-12.
[106]. Mueller, Mike (2017-08-01). "5 Things You Should Know about Geothermal Heat Pumps". Office of Energy Efficiency & Renewable Energy. US Department of Energy. Archived on 2021-04-15. Retrieved 2021-04-17.
[107]. "Dream or Reality? Electrification of the Chemical Process Industries". www.aiche-cep.com. Retrieved 2022-01-16.
[108]. Gillespie, Todd (2022-03-16). "Energy Costs Set to Reach Record 13% of Global GDP This Year". Bloomberg.com. Bloomberg. Retrieved 2022.
[109]. Scholten, D., Criekemans, D., & de Graaf, T. V. (2020). An Energy Transition Amidst Great Power Rivalry. Journal of International Affairs, 73(1), 195–203.
[110]. Chiu, Allyson; Guskin, Emily; Clement, Scott (2023-10-03). "Americans

don't hate living near solar and wind farms much as you might think". The Washington Post. Archived on 2023-10-03.

[111]. Nzaou-K., (2020). "Energy Transition Governance Research Materials". The Energy Transition Governance Research Materials.

[112]. Frumhoff, Peter C.; Heede, Richard; Oreskes, Naomi (2015-07-23). "The climate responsibilities of industrial carbon producers". Climatic Change. 132 (2): 157–171.

[113]. Mercure, J.-F.; et al. (2018). "Macroeconomic impact of stranded fossil fuel assets". Nature Climate Change. 8 (7): 588–593. ISSN 1758-678X. Archived on 2020-07-28. Retrieved 2020-08-19.

[114]. Howard, Emma (2015). "A Guide to Fossil Fuel Divestment". The Guardian. Archived on 2020-10-22. Retrieved 2020-03-29.

[115]. "Divestment Commitments". Fossil Free: Divestment. Archived on 2017-11-19. Retrieved 2020-03-29.

[116]. Geopolitics of the energy transition: Energy security. International Renewable Energy Agency (Report). Abu Dhabi. 2024-04-17. pp. 10–12. ISBN 978-92-9260-599-5. Retrieved 2024-04-17.

[117]. Hoppe, Thomas; et al. (2015). "Local Governments Supporting Local Energy Initiatives: Lessons from the Best Practices of Saerbeck (Germany) and Lochem (The Netherlands)". Sustainability. 7 (2): 1900–1931. ISSN 2071-1050.

[118]. Neves, Ana Rita; Leal, Vítor (December 2010). "Energy sustainability indicators for local energy planning: Review of current practices and derivation of a new framework". Renewable and Sustainable Energy Reviews. 14 (9): 2723–2735. ISSN 1364-0321.

[119]. SOVACOOL, Benjamin (2011). "Conceptualizing and measuring energy security: A synthesized approach". ink.library.smu.edu.sg. Archived on 2020-03-21. Retrieved 2020-03-29.

[120]. Strangleman, Tim (June 2001). "Networks, Place and Identities in Post-industrial Mining Communities". International Journal of Urban and Regional Research. 25 (2): 253–267. ISSN 0309-1317.

[121]. Bouzarovski, Stefan; et al. (2017). "Multiple transformations: theorizing energy vulnerability as a socio-spatial phenomenon". Geografiska Annaler: Series B, Human Geography. 99 (1): 20–41. ISSN 0435-3684.

[122]. "Training Available for Dislocated Coal Miners and Dependents « UMWA Career Centers, Inc". umwacc.com. 2015-11-04. Archived on 2020-03-29. Retrieved 2020-03-29.

[123]. Do, Thang Nam; Burke, Paul (2023). "Phasing out coal power in a developing country context: Insights from Vietnam". EnergyPolicy. 176

(May 2023 113512): 113512.
[124]. Marín, Anabel; Goya, Daniel (2021-12-01). "Mining—The dark side of the energy transition". Environmental Innovation and Societal Transitions. Celebrating a decade of EIST: What's next for transition studies?. 4. Korea Times. 2022-08-11. Retrieved 2023-01-03.
[125]. Pai, Sandeep; Carr-Wilson, Savannah (2018). Total Transition: The Human Side of the Renewable Energy Revolution. Rocky Mountain Books. ISBN 978-1-77160-248-8. on 2021-05-10. Retrieved 2020-10-03.
[126]. Stevis-Gridneff, Matina (2022-10-07). "Europe's Energy Crisis ExposesOld Fault Lines and New Power Dynamics". The New York Times.
[127]. Alova, G. (2020). "A global analysis of the progress and failure of electric utilities to adapt their portfolios of power-generation assets to the energy transition". Nature Energy. 5 (11): 920–927. Archived on 2021-03-21. Retrieved 2021-04-16.
[128]. Rand, Joseph; Hoen, Ben (July 2017). "Thirty years of North American wind energy acceptance research: What have we learned?". Energy Research & Social Science. 29: 135–148.
[129]. BP: Statistical Review of World Energy Archived 17 October 2020 at the Wayback Machine, Workbook (xlsx), London, 2016
[130]. World Energy Assessment Archived 12 November 2020 at the Wayback Machine (WEA). UNDP, United Nations Department of Economic and Social Affairs, World Energy Council, New York
[131]. "The Economics of Nuclear Power". World Nuclear Association. Retrieved 2024-12-06.
[132]. "Lazard Levelized Cost of Energy June 2024" (PDF). Lazard. Retrieved 2024-12-06.
[133]. "Statistical Review of World Energy (June 2018)". Retrieved 2019-09-27.
[134]. McGrath, Matt (2019). "China coal surge threatens Paris climate targets". Retrieved 2019-12-09.
[135]. Rosen, D.; Houser, Trevor (2007). "China Energy A Guide for Perplexed". piie.com. Retrieved 2020-04-25.
[136]. Zhou, Weihuan; Crochet, Victor; Wang, Haoxue (2025). "Demystifying China's Critical Minerals Strategies: Rethinking 'De-risking' Supply Chains". World Trade Review: 1–25.
[137]. "Renewable Energy in India - Indian Power Industry Investment". www.investindia.gov.in. Retrieved 2022-12-06.
[138]. Fuller, Gary (2024-04-19). "Clean-up Indian coal-fired power plant 'could have saved 720,000 lives'". The Guardian. ISSN 0261-3077. Retrieved 2024-04-19.

[139]. "The Electric Vehicle (EV) sector in India to boost both the ec..." www.investindia.gov.in. Retrieved 2022-12-06.
[140]. Do, Thang Nam; et al. (2021-12-01). "Vietnam's solar and wind power success: Policy implications for the other ASEAN countries". Energy for Sustainable Development. 65: 1–11. ISSN 0973-0826.
[141]. Do, Thang Nam (May 2024). "Insider perspectives on Southeast Asia's clean energy transition". Asia & the Pacific Policy Studies. 11 (2). ISSN 2050-2680.
[142]. Do, Thang Nam (2023). "Phasing out coal power in a developing country context: Insights from Vietnam". Energy Policy. 176 (May 2023 113512): 113512.
[143]. Do, Thang Nam; Burke, Paul J.; Baldwin, Ken; Nguyen, T.C. (2020). "Underlying drivers and barriers for solar photovoltaics diffusion: The case of Vietnam". Energy Policy. 144 (September 2020): 111561.
[144]. Do, Thang Nam; Burke, Paul J. (2024-06-01). "Phasing out coal power in two major Southeast Asian thermal coal economies: Indonesia and Vietnam". Energy for Sustainable Development. 80: 101451.
[45]. Tamma, Paola; Schaart, Eline; Gurzu, Anca (2019-12-11). "Europe's Green Deal plan unveiled". POLITICO. Archived on 2020-11-28. Retrieved 2019-12-29.
[146]. Simon, Frédéric (2019-12-11). "EU Commission unveils 'European Green Deal': The key points". www.euractiv.com. Archived on 2020-12-15.
[147]. European Investment Bank (2022). Digitalisation in Europe 2021-2022: Evidence from the EIB Investment Survey. European Investment Bank. ISBN 978-92-861-5233-7.
[148]. "Germany's energy transformation Energiewende". The Economist. 2012-07-28. Archived on 2018-01-15. Retrieved 2013-03-06.
[149]. "Wo steht Deutschland bei der Energiewende". Die Bundesregierung informiert | Startseite (in German). 2023-07-12. Retrieved 2023-09-14.
[150]. "The cost of financing for renewable power". www.irena.org. 2023-05-03. Retrieved 2023-11-03.
[151]. Egli, Florian; Steffen, Bjarne; Schmidt, Tobias S. (December 2018). "A dynamic analysis of financing conditions for renewable energy technologies". Nature Energy. 3 (12): 1084–1092. ISSN 2058-7546.
[152]. Furman, Jason. "A Reconsideration of Fiscal Policy in the Era of Low Interest Rates".
[153]. "Germany's energy reform: Troubled turn". The Economist. 2013-02-09. Archived on 2013-03-04. Retrieved 2013-03-06.
[154]. The Energy of the Future: Fourth "Energy Transition" Monitoring Report —

Summary. Berlin, Germany: Federal Ministry for Economic Affairs and Energy (BMWi). November 2015. Archived on 2016-09-20. Retrieved 2016-06-09.
[155]. Energy strategy 2050 Archived 19 May 2017 at the Wayback Machine, Swiss Federal Office of Energy, Federal Department of Environment, Transport, Energy and Communications (page visited on 21 May 2017).
[156]. "The Ten Point Plan for a Green Industrial Revolution (HTML version)". GOV.UK. Retrieved 2022-06-01.
[157]. Drax Group. "Drax Electric Insights". Drax Electric Insights. Archived on 2020-10-10. Retrieved 2020-09-10.
[158]. "Reducing UK emissions: 2020 Progress Report to Parliament". Committee on Climate Change. Archived on 2020. Retrieved 2020-09-10.
[159]. "Decarbonisation of Heat". Energy Systems Catapult. Archived on 2020-09-18. Retrieved 2020-09-10.
[160]. "Office for Low Emission Vehicles". GOV.UK. 2019-06-04. Archived on 2020-09-11. Retrieved 2020-09-10.
[161]. "Land use: Policies for a Net Zero UK". Committee on Climate Change. Archived on 2020-09-22. Retrieved 2020-09-10.
[162]. Frangoul, Anmar (2020-02-18). "UK government announces millions in funding for 'low carbon' hydrogen production". CNBC. Archived on 2020-10-29. Retrieved 2020-09-10.
[163]. Boydell, Ranald (2020-06-22). "Why zero-carbon homes must lead the green recovery from COVID-19". Conversation. Archived on 2020-09-09. Retrieved 2020-09-10.
[164]. Penman, Hamish (2020-09-02). "Gulf between government ambition and ability to deliver green energy transition". The Courier. Archived on 2020-09-17. Retrieved 2020-09-10.
[165]. Grubb, Professor Michael (2020-09-08). "Why a deal on energy could break the Brexit logjam". www.euractiv.com. Archived on 2020-09-12. Retrieved 2020-09-10.
[166]. "Big oil need not apply: UK raises for UN climate summit sponsorship". Climate Home News. 2020-08-18. Archived on 2020-09-23. Retrieved 2020-09-10.

Chapter (2)
Energy Insight

2.1. Shape of the Energy Industry in 2040

How much can the world energy system change in 16 years? If we go back 16 years to **2008**, we have events like the birth of the iPhone, the financial crisis, and Barack Obama elected as US president. All events, which, while not quite feeling like 'yesterday', still seem recent enough. How much can change in a decade-and-a-half? **2040** is just 16 years away, and from an energy perspective, we will see fundamental change, states Remi Eriksen, CEO of DNV [1, 2].

That is a bold assertion, given that transition so far in **2024**, appears to be partly in stall mode. Record emissions from fossil energy last year are on track to move even higher this year. High oil and gas prices are fuelling an exploration surge, while many renewable projects are experiencing an increase in cost due to inflationary and supply-chain pressures. Geopolitical developments over the past two years have brought energy security into sharp focus with the disruption of energy supplies and price shocks for energy importers. Local sourcing of both energy and energy infrastructure is emerging as a prominent national objective.

However, amidst these short-term perturbations, the long lines of development are clear: the energy landscape will look very different by **2040**. There are three major themes to keep an eye on: solar PV; electrification; and fossil fuels. These are themes executives and boards should address now in order to position and diversify their operations to thrive in the energy transition. Should countries and regions choose to reinforce their energy and climate policies beyond our anticipated measures, the associated trends and consequences would intensify at an accelerated pace [3].

A world becoming more efficient–delivering more useful energy with less waste!

According to our latest Energy Transition Outlook (**2023**), electricity demand is set to grow by 60% between now and **2040**. The shift to the use of electricity as an energy carrier brings enormous system efficiencies that will fundamentally alter how humanity uses energy. As the world's population increases and the global economy expands, it's clear that humanity will need more energy services (heating, cooling, transport etc.). Yet, energy demand will grow at a rate much slower than economic growth and even level off in the 2040s. The key to understanding this paradox lies in distinguishing between 'useful energy' and 'final energy [4, 5].'

Useful energy is the energy output that serves a specific purpose, such as heating a stove, propelling a vehicle, lighting a room, or running a machine. In our forecast, useful energy grows by 50% between now and 2040. What will not be growing nearly so fast is so-called 'final energy' – i.e. the total amount of energy required to meet various needs, which, crucially, includes both useful as well as wasted energy. Final energy demand refers to the energy delivered to end-use sectors (e.g. transportation, buildings, or manufacturing) and includes losses, mostly in the form of heat. In our forecast, global final energy demand grows by just 10% before flattening after 2040.

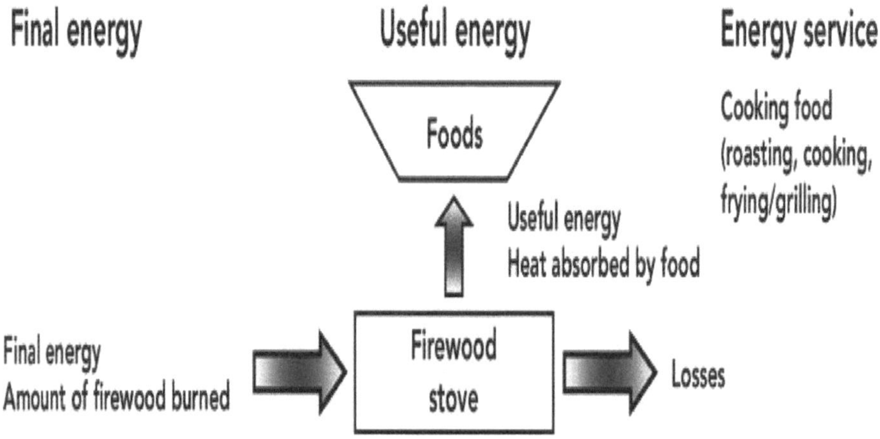

The difference between useful energy and final energy demand is clearly illustrated in the case of an internal combustion engine vehicle (ICEV), which is typically only 25% to 35% efficient – meaning that much of the chemical energy in the form of diesel or gasoline (the 'final energy demand') is wasted as heat, incomplete combustion, and friction before it provides the useful work of propelling the vehicle (useful energy). As ICEVs are replaced by electric vehicles (EVs) that have much higher efficiencies (>90%), enormous energy losses will be eliminated. As a consequence, as energy end use becomes increasingly electrified, there will be a lot less demand for oil [6].

2.2. Solar PV

2023 saw an almost 50% increase in the amount of new renewable projects coming online. Global annual renewable capacity additions amounted to 510 gigawatts (GW) in 2023, with solar PV alone accounting for three-quarters of the total additions. This is the fastest rate of growth of renewable capacity additions in the past two decades and the growth will continue. A few regions stand out, like Europe, the United States and Brazil which hit all-time highs, whereas China's acceleration was extraordinary - which we have detailed in a recent report, DNV Energy Transition Outlook China 2050 (released April 2024) [7, 8].

By 2040, we expect solar PV to have an installed capacity of 9.7 TW and produce almost 14 TWh which is 30% of all electricity generation. Solar PV will be the cheapest new electricity source globally, with exceptions only in areas with less conducive irradiation conditions like the higher northern latitudes. By 2040 the reduction in cost per kWh will be another 40% compared with today. Speed of deployment at utility scale as well as ample industrial capacity for producing panels paves the way for solar PV dominance as the major source of electricity generation [9].

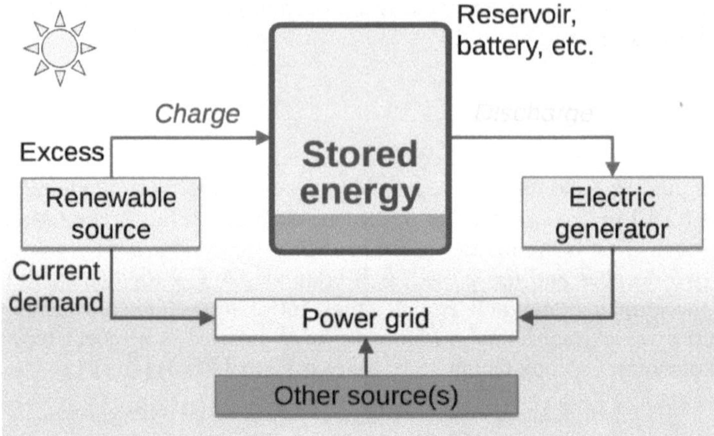

With variable renewable energy, there will be a need for flexibility and storage in a growing electricity grid. This is where Li-ion batteries play the primary role. These batteries will either be integrated with renewables or operate as standalone systems. The ability to store excess energy during peak sunlight hours and then sell it when after-dark prices rise gives solar + storage an edge over standalone solar PV and provides a solution to the variability problem. This means that existing thermal power plants will increasingly operate with (rather than instead of) renewables, with the viability of those thermal sources increasingly tied to their flexibility to follow demand. We expect 6.2 terawatt-hours (TWh) of lithium-ion battery storage by 2040, with a substantial part integrated with renewable energy sources and a matching amount of storage available from EVs enabled to dispatch stored electricity to the grid [9].

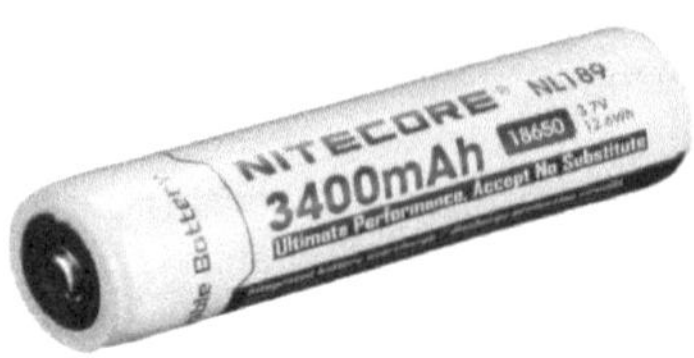

2.3. Electrification

Technology, cost, and policy factors are all driving increased electricity use, which naturally holds efficiency advantages over other energy carriers. From a decarbonization perspective, everything that can be electrified should be electrified. Because electricity is fairly easy to decarbonize compared with other energy carriers, ever-more ambitious decarbonization policies will inevitably favour electricity. Additionally, the costs of solar and wind power are expected to continue decreasing (global average by 45% mid-century), making electricity increasingly more cost-effective compared with other fuels, especially hydro-carbons which will increasingly be subject to carbon tax in most jurisdictions [10].

The delivered cost of electricity is of course highly dependent on market designs and on the manner in which grid build-out is supported and funded. New applications requiring modern energy are also emerging — for example, AI, communication appliances, and air conditioning — for which there are few or no alternatives to electricity.

FIGURE 2

Share of electricity in final energy demand

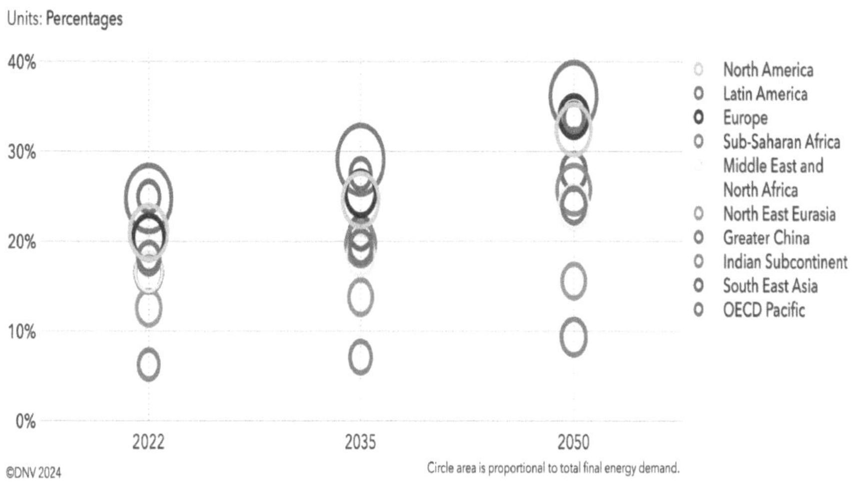

Electricity stands centrally in the ongoing global energy transition, shaping innovation and strategies in both supply and demand sectors, with China leading the way owing to its focus on achieving energy independence (Figure). By 2040, electricity will account for 28.4% of the world's total energy demand, a significant increase from 20% in 2023, but more importantly guidance for how the future beyond 2040 will unfold. Renewable electricity generation will power much of this growth, increasing from 9.2 petawatt-hours per year in 2023 to 32.5 petawatt-hours by 2040. This increase means that the share of renewables in electricity generation will rise from 31% 2023 to 67% by 2040 (Figure) [11].
\

The emergence of so-called 'new' power systems (i.e. those characterised by a high penetration of renewable sources) is a reality that deserves a great deal more attention than it is currently given. Nations stand to benefit greatly in competetiveness and security of their national energy systems if issues such as adequacy, flexibility, grid capacity, and power market design are tackled early. See, for example, DNV's report on New Power Systems (May 2024).

At present, there is a shortage of grid capacity to accommodate the coming expansion of both generation and load (i.e. demand). While enhanced transmission planning and build-out is the long-term solution to addressing a capacity-constrained electric grid, there are near-term tools at our disposal to ensure we are getting the most out of the grid we have. Their deployment does not require new gridlines and are becoming increasingly standardized and adopted in regulatory frameworks. These 'non-wire' grid enhancing technologies offer a promising path-way to rapid and cost-effectively increase grid capacity and efficiency, facilitating the integration of renewable energy and enhancing grid reliability over a bridging period to the 2030s.

From a financial perspective, an implementation of this grid enhancement approach not only reduces the capital expenditure associated but also lowers operational costs by enabling an earlier penetration of renewable energy sources, which are typically cheaper to operate than fossil-fuel-based sources.

2.4. Oil and Gas

Historically, oil demand has increased roughly 1% annually due to population and economic growth, maintaining a significant presence in the transportation sector where it has comprised about 70% of the energy use since 2017. However, this trend is set to shift significantly as EV adoption reaches double digit percent-tages of new vehicles sales in several major markets.

Our forecasts show that by the late 2020s, EVs will make up 50% of new passenger vehicle sales in Greater China and Europe. This shift will extend to the OECD Pacific and North America by the early 2030s, and it is expected to reach a market share of 50% by 2031 globally, which means that by 2040 almost 40% of the passenger vehicle fleet will be electric.

The impact of EVs is particularly pronounced in China, the world's largest passenger vehicle market. In 2023, EVs made up about 20% of new passenger vehicle sales, rising to about 25% in recent months of 2024. Mid-2023, the Chinese oil giant Sinopec unexpectedly announced that it expects gasoline demand in China to peak in 2023, two years earlier than was set out in previous forecasts. Globally, we expect oil demand to peak before the end of this decade, with its subsequent decline primarily driven by rapid EV uptake, steepening in the 2030s as other key transport sectors like shipping and aviation start their decarbonization journeys deploying alternative fuels such as ammonia, methanol and other syn-thetic fuels.

Looking ahead, we forecast that by 2040, oil demand in China's road sector will decrease by 75%, which will have major repercussions for both domestic oil production and the global oil market. By 2040 the global demand for oil will have dropped 22% from today's levels and will be in perpetual decline.

Natural gas demand vary by region, typically growing in low- and medium-income regions and declining in OECD regions (Figure). Furthermore, there will be demand for natural gas in new sectors, particularly with increasing use in maritime transport, and as a feedstock for making blue hydrogen and ammonia. Over the past 10 years, power generation accounted for 35% of gas use. This share will rise to 37% in 2026 before receding to 30% by 2040, a shift driven by the expansion of renewable energy sources. However, global demand for natural gas increases to 5,035 billion m3 around 2027 and plateaus for about a decade before gradually declining to 4,800 by 2040 billion m3 in 2040.

Gas consumption in Europe and the OECD Pacific is projected to decrease to approximately 55% and 40% of 2023 levels by 2040, respectively. This significant reduction is largely due to the natural gas supply constraints following Russia's invasion of Ukraine. In Greater China, gas demand is expected to increase by 28% by 2030, but it will level off as it approaches 2040. Both Greater China and the

Indian Subcontinent are characterized by high demand and limited domestic natural gas resources and will import larger amounts via LNG. In contrast, gas production in North America and Europe is projected to decrease by half, which corresponds to a 10% reduction in their market shares.

The Middle East and North Africa, along with Russia, have long been net exporters of gas and are anticipated to continue in this role. Notably, export volumes from the Middle East and North Africa are expected to double by 2040, with its market share also rising from todays 10% to reach 35% in 2040.

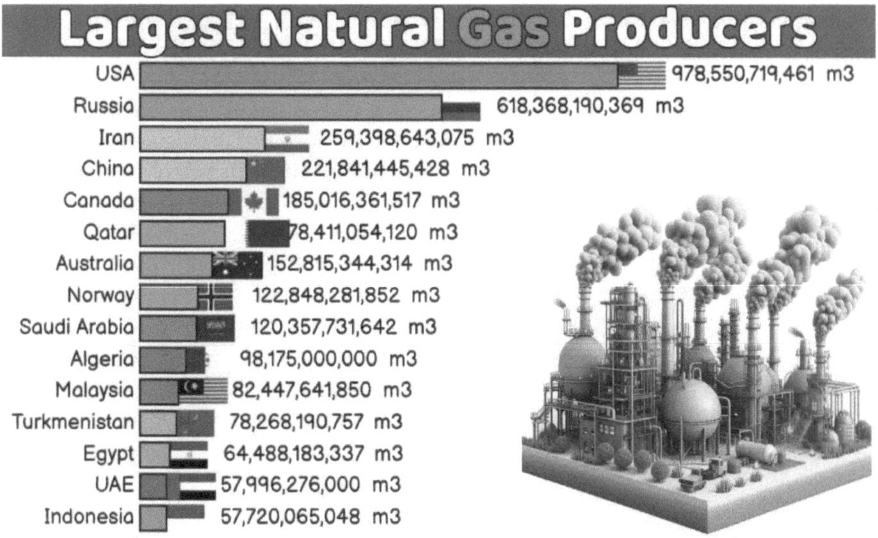

2.5. Future Projections

In 2023, electricity generation from wind and solar sources was projected to exceed 30% by 2030 [12-15].

A December 2022 report by the IEA forecasts that over 2022-2027, renewables are seen growing by almost 2,400 GW in its main forecast, equal to the entire installed power capacity of China in 2021. This is an 85% acceleration from the previous five years, and almost 30% higher than what the IEA forecast in its 2021 report, making its largest ever upward revision. Renewables are set to account for over 90% of global electricity capacity expansion over the forecast period. To achieve net zero emissions by **2050**, IEA believes that 90% of global electricity generation will need to be produced from renewable sources [16].

In June **2022**, IEA Executive Director Fatih Birol said that countries should invest more in renewables to "ease the pressure on consumers from high fossil fuel prices, make our energy systems more secure, and get the world on track to reach our climate goals [17]."

China's five year plan to **2025** includes increasing direct heating by renewables such as geothermal and solar thermal.

After a transitional period, renewable energy production is expected to make up most of the world's energy production. In **2018**, the risk management firm, DNV GL, forecasts that the world's primary energy mix will be split equally between fossil and non-fossil sources by **2050**.

Middle Eastern nations are also planning on reducing their reliance fossil fuel. Many planned green projects will contribute in 26% of energy supply for the region by **2050** achieving emission reductions equal to 1.1 Gt CO_2/year [18].

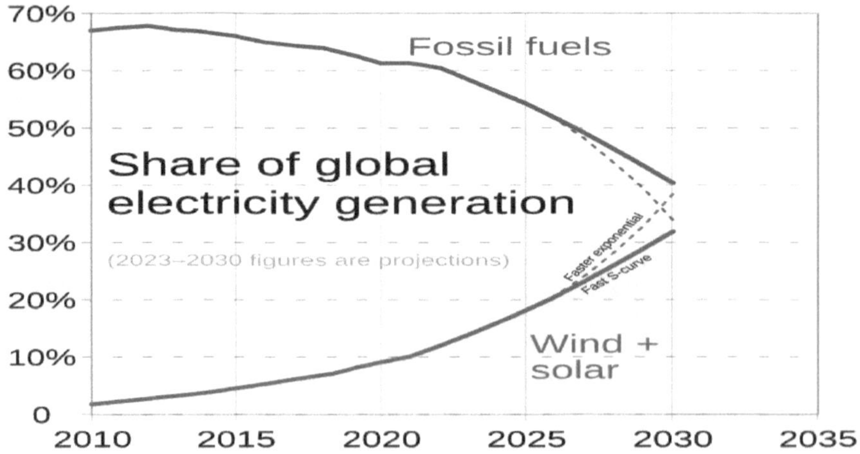

2.5. References

[1]. Armaroli, Nicola; Balzani, Vincenzo (2011). "Towards an electricity-powered world". Energy and Environmental Science. 4 (9): 3193–3222. ISSN 1754-5692.

[2]. Armaroli, Nicola; Balzani, Vincenzo (2016). "Solar Electricity and Solar Fuels: Status and Perspectives in the Context of the Energy Transition". Chemistry – A European Journal. 22 (1): 32–57.

[3]. "Global renewable energy trends". Deloitte Insights. Archived from on 29 January 2019. Retrieved 28 January 2019.

[4]. "Renewable Energy Now Accounts for a Third of Global Power Capacity". irena.org. 2 April 2019. Archived on 2 April 2019. Retrieved 2 Dec. 2020.

[5]. "2023 Levelized Cost Of Energy+". www.lazard.com. Retrieved 10 June 2024.

[6]. IEA (2020). Renewables 2020 Analysis and forecast to 2025. p. 12. Archived on 26 April 2021. Retrieved 27 April 2021.

[7]. "Renewables 2022". Global Status Report (renewable energies): 44. 14 June 2019. Retrieved 5 September 2022.

[8]. "Global Electricity Review 2025". Ember. Retrieved 11 April 2025.

[9]. "Renewables - Energy System". IEA.

[10]. Ritchie, Hannah; Roser, Max; Rosado, Pablo (2024). "Renewable Energy". Our World in Data.

[11]. Heidari, Negin; Pearce, Joshua M. (2016). "A Review of Greenhouse Gas Emission Liabilities as the Value of Renewable Energy for Mitigating Lawsuits for Climate Change Related Damages". Renewable and Sustainable Energy Reviews. 55C: 899–908. Archived on 28 July 2020. Retrieved 26 February 2016.

[12]. "Global Trends in Renewable Energy Investment 2020". Capacity4dev / European Commission. Frankfurt School-UNEP Collaborating Centre for Climate & Sustainable Energy Finance; BloombergNEF. 2020. Archived on 11 May 2021. Retrieved 16 February 2021.

[13]. Ritchie, Hannah; Roser, Max; Rosado, Pablo (27 October 2022). "Energy". Our World in Data.

[14]. Bond, Kingsmill; et al. (2023). "Report / 2023 / X-Change: Electricity / On track for disruption". Rocky Mountain Institute. Archived on 13 July 2023.

[15]. "Record clean energy spending is set to help global energy investment grow by 8% in 2022 - News". IEA. 22 June 2022. Retrieved 27 June 2022.

[16]. "China's New Plan for Renewable Energy Development Focuses on Consumption". www.fitchratings.com. Retrieved 27 June 2022.

[17]. Claeys, Bram; Rosenow, Jan; Anderson, Megan (27 June 2022). "REPower EUight energy policy recipe to move away from Russian gas?". www.euractiv.com. Retrieved 27 June 2022.
[18]. Gan, Kai Ernn; et al. (2023). "Enhancing Renewable Energy Systems, Contributing to Sustainable Development Goals of United Nation and Building Resilience against Climate Change Impacts". Energy Technology. 11 (11). ISSN 2194-4288.

Chapter (3)
Energy Storage

3.1. Preface

Energy storage is a crucial technology to provide the necessary flexibility, stability, and reliability for the energy system of the future. System flexibility is particularly needed in the EU's electricity system, where the share of renewable energy is estimated to reach around 69% by 2030 and 80% by 2050.

The Commission adopted in March 2023 a list of recommendations to ensure greater deployment of energy storage, accompanied by a staff working document, providing an outlook of the EU's current regulatory, market, and financing framework for storage and identifies barriers, opportunities and best practices for its development and deployment.

3.2. Key Facts

System flexibility is particularly needed in the EU's electricity system, where the share of renewable energy is estimated to reach around 69% by 2030 and 80% by 2050 (from 37% in 2021).

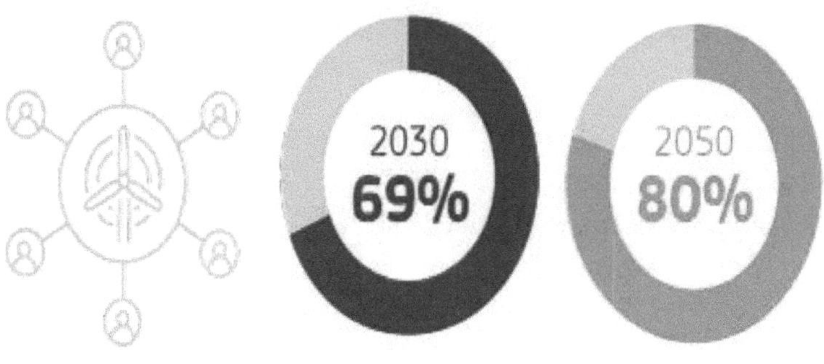

The need for flexibility in the electricity system will increase significantly in all EU countries, reaching 24% (288 TWh) of total EU electricity demand in 2030 and 30% (2 189 TWh) by 2050 across all timescales (from 11% in 2021).

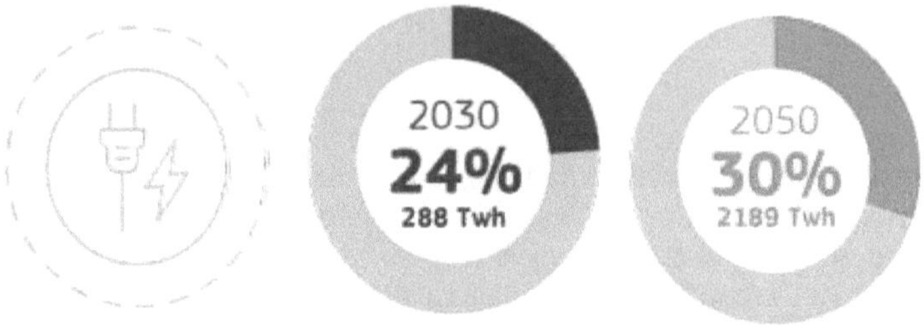

The daily, weekly and monthly flexibility requirements should reach averages of 2.52 TWh/day, 14.6 TWh/week and 41.68 TWh/month by **2050.**

On average across the EU, the overall flexibility requirements increase signify-cantly when the share of renewable generation in the electricity system is above 74% of the total installed capacity.

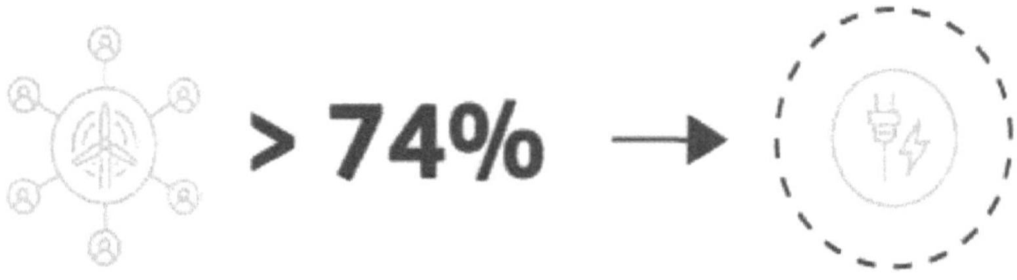

Many European energy-storage markets are growing strongly, with 2.8 GW (3.3 GWh) of utility-scale energy storage newly deployed in **2022**, giving an estimated total of more than 9 GWh.

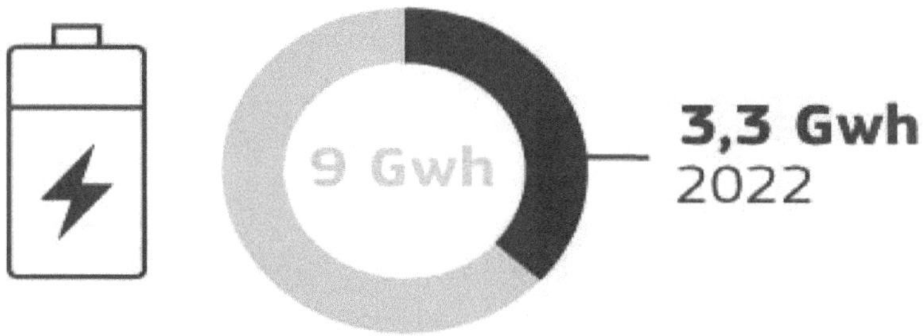

Looking forward, the International Energy Agency (IEA) expects global install-led storage capacity to expand by 56% in the next 5 years to reach over 270 GW by **2026**.

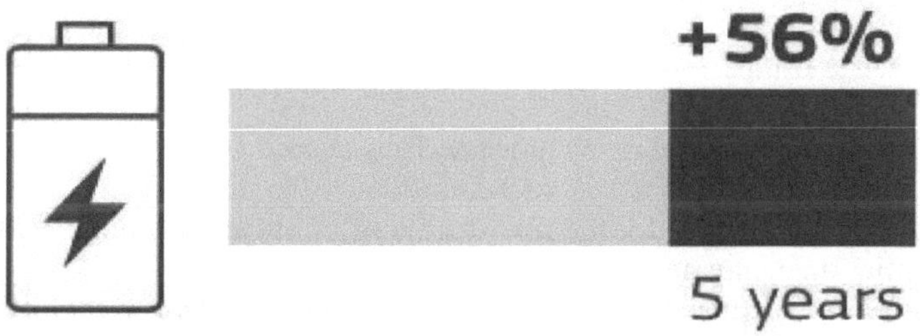

Different studies have analyzed the likely future paths for the deployment of energy storage in the EU. These studies point to more than 200 GW and 600 GW of energy storage capacity by **2030 and 2050** respectively (from roughly 60 GW in **2022**, mainly in the form of pumped hydro storage).

3.3. Key Actions

The EU needs a strong, sustainable, and resilient industrial value chain for energy-storage technologies. There is an increasing demand for data transparency and availability, and greater data granularity, including network congestion, renewable energy curtailment, market prices, renewable energy, greenhouse gas emissions content and installed energy-storage facilities. This need becomes more important for decisions about investing in, choosing a location for, and evaluating new energy-storage facilities.

EU countries are essential in the process. They should:

- assess capacity needs for the relevant energy storage technologies as well as potential financing gaps
- identify actions necessary to remove barriers to the deployment of demand response and behind-the-meter storage
- establish cost-effective processes, such as competitive bidding systems and suitably designed capacity mechanisms
- accelerate the deployment of storage facilities and other flexibility tools in islands, remote areas and the EU's outermost regions with insufficient or unstable grid capacity
- publish detailed data on the energy market to facilitate investment decisions on new energy storage facilities
- support research and innovation - in particular, long-term energy storage – and launch technology accelerator programs and dedicated support sche-mes

EU initiatives can play a role in:

- facilitating the deployment of flexible resources to complement intermittent renewable production

- addressing specific barriers for distributed flexibility sources, including energy storage

Documents

- 14 MARCH 2023

Staff working document on the energy storage - underpinning a de-carbonized and secure EU energy system

- Commission recommendation on energy storage – underpinning a decarbonized and secure EU energy system (C/2023/1729)
- News: Commission recommendations on how to exploit the potential of energy storage (14 March 2023)
- Energy system integration

3.4. Energy storage Systems Towards 2050

3.4.1. Introduction

The environmental impact of fossil fuels [1-5], their fluctuation in price, and their limited resources urged the scientists to find sustainable energy sources with low or no environmental impacts. Different strategies have been applied, such as improving the current processes' efficiency through waste heat recovery [6-8], developing environmentally friendly energy-efficient devices [9-13], and applying renewable energy sources. Renewable energy sources are sustainable with low or no environmental impacts [14, 15]; therefore, they can potentially replace fossil fuels soon. During the last decade, there is a significant decrease in the price of renewable energy sources where the cost of the PV decreased by 90%, and wind energy fell by 50% [16]. This decrease in the price was accompanied by a significant increase in renewable energy capacity, which urged finding proper renewable energy storage to accommodate such an increase in the renewable energy sources' capacity. Therefore, energy storage systems have become one of the most important research topics due to maintaining energy production from renewables. As described in Refs. [17], the main Energy storage techniques can be classified into mechanical, electrochemical, thermal, electrical, and magnetic [1].

Figure highlights the main criteria that can guide the proper selection of different renewable energy storage systems. Various criteria can help decide the proper energy storage system for definite renewable energy sources, as shown in the figure. For instance, solar energy and wind energy are high intermittences daily or seasonally, respectively, compared with geothermal energy. Moreover, the integration of the different hybrid renewable energy sources can minimize such intermittency. Such availability/intermittency of renewable energy sources and the energy capacity can decide the type and size of the energy storage system. Application is also crucial in selecting the proper energy storage system.

For instance, industrial, domestic, or transportation requires different energy storage systems. For example, in the industrial application case, the energy required is constant with time, so no problem with the peak power consumption. While there is a peak power in the domestic case, and there is an annual change in the energy demand according to the population growth. Meanwhile, in trans-portation, energy consumption has several peak points, such as in the morning, noon, and evening. A proper energy storage system must satisfy the requirements according to the application. The available technology plays a main factor in deci-ding the appropriate energy storage system. The mature energy storage technology will have different sizes of the system that can accommodate varying energy capa-cities with reasonable cost and lifetime. Also, mature technology will grantee a reliable energy storage system with rapid response and low problems. Infra-structure refers to the available facilities, including the distance from the renewable energy sources, the available land area for implementing the energy storage system, and the weather's geographical suitability for the proposed energy storage system. For instance, pumped hydro energy storage systems require a large area and suitable for large applications, i.e., big capacities, while batteries require mode-rate temperature to extend their lifetime and suitable for small capacities. The available grid connection can reduce the renewable energy storage system load where this grid will be the main energy storage. Therefore only an auxiliary energy storage system will be used in the low energy demand or under any disturbance in the grid. Although renewable energy storage systems have small environmental impacts, it can be considered a factor in deciding the best renewable energy storage systems. For example, battery disposal can negatively affect the environment as it contains some hazardous materials, while hydro pump storage has visual pollution and requires a large area of the land.

This prospect describes research work presented at the 11th Conference on Sustainable Energy and Environmental Protection SEEP2018, which was held between 8th and 11th May 2018 at the University of the West of Scotland, Paisley, UK.

During this conference, the following topics were presented: renewable energy (wind, solar, marine, geothermal, biomass, ...etc), energy storage systems, energy efficiency, waste energy recovery, energy management, and various articles related to the conference scope.

More than 250 delegates have attended the conference and more than 350 papers have been presented during this conference, around 80 papers were invited to submit an extended version to Energy, and 46 papers were accepted for publications. In this prospect, the conference theme on energy storage systems toward 2050 was described; also the paper includes a brief description of the 46 accepted papers.

3.5. References

[1]. A.G. Olabi. Renewable energy and energy storage systems Energy (2017).

[2]. Z. Zhang et al. Status and perspective of CO2 absorption process Energy (2020).

[3]. K. Elsaid et al. Environmental impact of emerging desalination technologies: a preliminary evaluation. J Environ Chem Eng (2020).

[4]. K. Elsaid et al. Environmental impact of desalination technologies: a review. Sci Total Environ (2020).

[5]. A.G. Olabi et al. Waste heat-driven desalination systems: Perspective Energy (2020).

[6]. K. Elsaid et al. Recent progress on the utilization of waste heat for desalination: a review. Energy Convers Manage (2020).

[7]. H. Jouhara et al. Editorial: industrial waste heat recovery. Energy (2018).

[8]. M.A. Abdelkareem et al. Significance of diffusion layers on the performance of liquid and vapor feed passive direct methanol fuel cells. Energy (2020)

[9]. A.G. Olabi et al. Application of graphene in energy storage device – a review. Renew Sustain Energy Rev. (2021).

[10]. A.H. Alami et al. Titanium dioxide-coated nickel foam photoelectrodes for direct urea fuel cell applications. Energy (2020).

[11]. Battery energy storage systems and SWOT (strengths, weakness, opportunities, and threats) analysis of batteries in power transmission. 2022, Energy.

[12]. Battery thermal management systems: Recent progress and challenges 2022, International Journal of Thermofluids.

[13]. Luisa F. Cabeza. Energy storage for grid-scale applications: Technology

review and economic feasibility analysis. Renewable Energy, Volume 163, 2021, pp. 1754-1772
[14]. Guido Francesco Frate, Umberto Desideri. Overview on recent developments in energy storage: Mechanical, electrochemical and hydrogen technologies. Energy Conversion and Management, Volume 132, 2017, pp. 372-387.

Chapter (4)
Mobility Transition

4.1. Preface

Hermann Knoflacher has been criticizing auto cities and car dependency for decades. With his walking gear, he caricatures the enormous spatial demands of motorized private transport (2007).

Mobility transition [1, 2] is a set of social, technological and political processes of converting traffic (including freight transport) and mobility to sustainable transport with renewable energy resources, and an integration of several different modes oprivate transport and local public transport. It also includes social change, a redistribution of public spaces [3], and different ways of financing and spending money in urban planning. The main motivation for mobility transition is the reduction of the harm and damage that traffic causes to people (mostly but not solely due to collisions) and the environment (which also often directly or indirectly affects people) in order to make (urban) society more livable, as well as solving various interconnected logistical, social, economic and energy issues and inefficiencies.

4.2. Motivation

4.2.1. Environmental Damage

An important goal is the reduction of greenhouse gas emissions such as CO2. To achieve the goal set in the Paris Agreement, that is, to restrict global warming to clearly below 2 °C, the burning of fossil fuels is to be discontinued around

2040 [4]. Because the CO2 emissions of traffic practically need to be reduced to zero [5], the measures taken so far in the transport sector are not sufficient in order to achieve the climate change mitigation goals that have been set [6].

4.2.2. Air Pollution

A mobility transition also serves health purposes in the metropolitan regions and large cities and is intended in particular to counteract the massive air pollution [7]. For example, in Germany in 2015, traffic caused about 38% of human-related nitrogen oxide emissions [8]. According to Lelieveld et al. (2015), air pollution from land traffic alone killed around 164,000 people in 2010; in Germany alone, it was over 6,900 people [9]. A 2017 study by the same lead author concluded that air pollution from road traffic in Germany causes 11,000 deaths every year that could potentially be avoided. This figure is 3.5 times the number of fatalities from accidents [10].

Classic urban mobility using the example of Aachen: privately owned automobiles characterize the streetscape.

To demonstrate how much road traffic contributes to air pollution in Germany, for every 100 inhabitants, 58 of them owned passenger cars, according to Federal Statistical Office of Germany [11].

4.2.3. Accident Fatalities, Quality of Life, Aggressive Behaviour

Further motives for the mobility transition are the desire for less noise, streets with quality of life and lower accident risks. According to estimates by the European Environment Agency, 113 million people in Europe are affected by road noise at unhealthy levels [12]. With increasing traffic and commuter numbers, many citizens also wished for more attractive places to spend time in public spaces [13]. A mobility transition therefore also serves to increase the quality of life [14].

The mobility transition is also seen by some as a means of reducing aggressive behaviour in traffic (road rage) and in society [15]. Studies indicate that people in large and expensive cars are more likely to behave recklessly [16]. According to the German Verkehrsklima 2020 (Traffic Mood 2020) study, women feel more insecure in traffic than men, and they want more controls and stricter laws [17]. On the other hand, the "evil eye" design of vehicles is increasingly used by manufacturers to sell vehicles to drivers who want to feel strong and superior on the road [18]. Accident reporting by the press and the police sometimes paints a distorted picture [19].

4.2.4. Traffic Congestion

Traffic congestion has been increasing in streets and roads. Traditional traffic policy usually relies on expanding the roads to solve the congestion problem [20]. Worldwide, the main causes are urbanization and the purchase of more automobiles as prosperity increases. A return to more public and non-motorized transport is likely in the future [21].

4.2.5. Peak Oil

Petroleum production is approaching its peak, or by some estimates may already have been passed in the 2020s [22]. The Earth's oil reserves are finite, and oil extraction will become inadequate to power as many petroleum-fueled vehicles. Sooner or later in the 21st century, mobility must rely on other energy sources [22].

4.3. Mobility Transition Concept

4.3.1. Origins

Action group "Stop the Child Murder" protesting in front of the Dutch Parliament in 1972. The banners state "Stop child murder" and "Safe walking and cycling paths".

There has been criticism of automotive cities and car dependency since at least the 1960s. In the Netherlands, Provo Luud Schimmelpennink's 1965 White Bicycle Plan was an early attempt to stop the rising death toll due to car-related traffic accidents, and to stimulate cycling as a safer and healthier alternative for short-distance travel in the city of Amsterdam [23]. Although the plan itself was a complete failure, it drew widespread publicity and influenced urban planning ideas around the world – with the white bicycle becoming 'an almost mythical worldwide symbol for a better world' [23]. It inspired the emergence of both strongly anti-car movements such as Kabouter (Gnome), Amsterdam Autovrij ("Amsterdam Car-Free") and De Lastige Amsterdammer ("The Troubled/ Trouble-some Amsterdammer"), as well as pro-cycling movements in Amsterdam and elsewhere in the Netherlands in the early 1970s [23].

A prominent example was protest group Stop de Kindermoord ("Stop the Child Murder"), founded in 1972 (formalised in 1973) by a journalist from Eindhoven whose young daughter was killed in a traffic accident, and shortly thereafter another daughter of his was almost killed as well [23]. The movement highlighted how lethally dangerous traffic had become for children in particular, and that the authorities had failed to acknowledge and address the problem [23]. It mobilised parents, teachers, journalists, other citizens and politicians; even right-wing politicians, who had traditionally promoted automobile interests, were influenced by the campaign and became more willing to adopt preventive measures [23].

In Autokind vs. Mankind (1971) and on Nature of Cities (1979), American author Kenneth R. Schneider vehemently criticized the excesses of automobile dependence and called for a struggle to halt and partially reverse negative developments in transportation, although he was largely ignored at the time [24].

An early theorist on mobility transitions was American cultural geographer Wilbur Zelinsky, whose 1971 paper "The Hypothesis of the Mobility Transition" formed the basis of what has become known as the Zelinsky Model [25, 26]. In 1975, Austrian civil engineer and transportation planner Hermann Knoflacher sought to promote cycling traffic in Vienna. He caricatured the enormous spatial demands of automobiles with his self-invented Gehzeug ("walking gear/vehicle") [27, 28].

4.4. Definitions and Scope

Cycling in Copenhagen also means organizing commuter traffic by bike.
45 per cent of the population commutes by bicycle.

German dictionary Duden define 'mobility transition' (Verkehrswende) as "fundamental conversion of public transport [especially with ecological objectives]" (German: „grundlegende Umstellung des öffentlichen Verkehrs [besonders mit ökologischen Zielvorstellungen]") [29]. Adey et al. (2021) defined 'mobility transition' as 'the necessary and inevitable transformation from a world in which mobility is dominated by the use of fossil fuels, the production of greenhouse gases and the dominance of automobility to one in which mobility entails reduced or eliminated fossil fuels and GHG emissions and is less dependent on the automobile '[1].

According to a 2016 thesis paper by Agora Verkehrswende – a joint initiative of Stiftung Mercator [30] and the European Climate Foundation – the goal of a traffic transition (Verkehrswende) in Germany is ensuring climate neutrality in transport by 2050 [20]. It must be based on two pillars:

- Mobility transition (Mobilitätswende): The goal is a significant reduction of energy consumption. The mobility transition is intended to bring about a qualitative change in traffic behaviour (Verkehrsverhalten), in particular avoiding and relocating traffic. An efficient design of the traffic systems without restricting mobility should be achieved [20].

- Energy transition in traffic (Energiewende im Verkehr, see also phase-out of fossil fuel vehicles): In order to decarbonise traffic, the conversion of the energy supply of traffic towards renewable energy is considered a necessity [31].

A mobility transition also includes a cultural change, in particular a re-evaluation of "the street". Currently, the primary purpose of streets is to direct traffic through the city with as little disruption as possible. In the future, the dominance of the car should give way to equal rights for all modes of transport [32].

In an expanded definition, the mobility transition is distinguished from a pure propulsion transition on the one hand to a fundamental mobility transition on the other [33]:

- Propulsion transition (Antriebswende): the gradual replacement of internal combustion engines by those powered by hydrogen, fuel cells or battery-electric power [33, 34].

- Traffic transition (Verkehrswende): private car traffic [35] is reduced or replaced by other modes of transportation. In the large cities and metropolitan regions in particular, the focus is increasingly on establishing and spreading alternative means of transport - from the expansion of public transport to the promotion of so-called active transport (pedestrian and bicycle traffic), the approval of new electrified micro-vehicles such as e-scooters and the range of different mobility services (the so-called MaaS, "mobility as a service") [33].

- Mobility transition (Mobilitätswende): This perspective takes into account not only the distances travelled and the means of transport used for them, but also the socio-economic, cultural and spatial dynamics and constraints that cause the need to overcome distances. These include, for example, settlement and transport policies, housing and labour markets, social policy and migration. The need to quickly overcome distances is not understood as an invariant characteristic of people, but as part and prerequisite of the current, growth-oriented capitalist shape of society [33].

Electric cars in a carsharing as part of mixed-mode commuting
(San Francisco City Hall 2009).

In some cases, a mobility transition is also presented as a paradigm shift of the 'understanding of ownership'. Collective use of means of transport makes it possible to use modes of transportation 'adapted to specific needs', such as car-sharing, peer-to-peer car-sharing, bicycle-sharing systems. It also enables connec-ting different modes of transportation to one another on a route to be travelled. Electromobiles could better exploit their advantages in networking with other means of transport. Electric vehicles adapted to the respective uses can be small or large depending on the application, and do not (always) have to be designed for long distances. A suitable charging infrastructure is required. Under certain circumstances, in such an environment it will no longer be necessary to own pri-vate transport for one's own use [36].

In Germany, the mobility transition can be contrasted to the Bundesverke-hrswegeplan 2030 ('Federal Transport Routes Plan 2030'). The mobility transition is based on avoiding traffic and shifting to rail, but the Bundesverkehrswegeplan is based on the construction and expansion of trunk roads in Germany (including but not limited to the Autobahn) [37, 38]. Transport scientist Heiner Monheim [de] regards the transition as a "turning away from car subsidies through billions of euros in road network expansion". He sees a decisive change in the priorities of transport policy as a necessary condition to achieve this [39].

The Umweltbundesamt announced that in 2018, the sum of all environmentally harmful subsidies in Germany was 65.4 billion euros, almost half of them in the areas of traffic and transport. In traffic, such subsidies with harmful effects even increased from 2012 to 2018 [40].

4.5. Changes in Behaviour due to the COVID-19 Pandemic

During the pandemic, the efficiency of urban land use has been reconsidered. (Temporary) outdoor areas for restaurants are sometimes more economical than parking spaces, and enhance the cityscape [41].

The COVID-19 pandemic made it clear that work and transport can be organized differently, even in a comparatively short time. An increased focus on working from home could save millions of tonnes of greenhouse gases [42-44].

4.5.1. Measures in Passenger Transport

4.5.1.1. Overview

Bike freeway F35 in the Netherlands (Enschede)

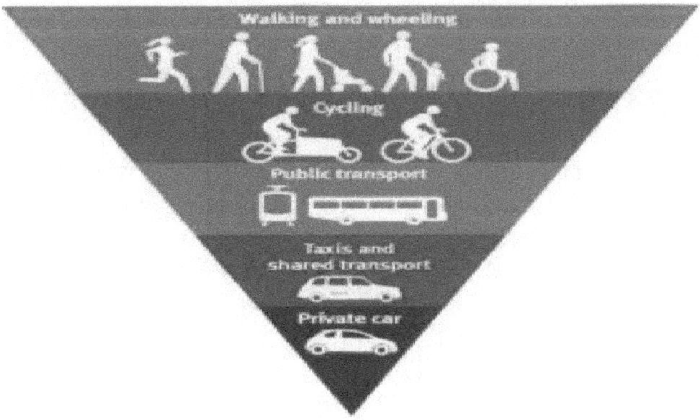

Mobility pyramid

Various measures have been proposed by different people and groups to achi-eve a mobility transition.

In a 2017 position paper, German think tank Agora Verkehrswende described how a climate-neutral conversion of transport would be possible by 2050 without sacrificing mobility. In addition to technological innovations, there are new traffic concepts, regulatory measures and cultural change. Multi-link transport chains (Intermodal passenger transport) are considered [45]. Amongst other things, there were also studies on this in 2019 by the Verkehrsclub Deutschland [de] (VCD, "Traffic Club Germany") and the Heinrich Böll Foundation [46].

4.5.1.2. Mobility Transition

Various measures have been proposed to achieve the mobility transition – in particular a significant reduction in energy requirements and a change in traffic behaviour:

The Autonomous Bus Vienna (2019–2021) was an experimental company for self-driving buses.

Major changes can succeed with the help of traffic avoidance, and a shift tow-ards sustainable transport in the form of pedestrian traffic, cycling, rail trans-port and local public transport [47, 48]. According to a 2010 report, each person in Germany in 2008 conducted an average of 3.4 trips a day, with an average length of 11.5 kilometers. On average, private cars were parked for around 22.5 hours a day, because they were used for only 1 hour and 19 to 28 minutes a day [49]. Electric cars with short range, bicycles, electric bicycles (e-bikes), pedicels, cargo bikes, but also recently e-scooters, are usually well suited for a majority of these routes. The joint use of automobiles in car-sharing could increase the utilization of the vehicles and lead to fewer cars being needed overall. This could also reduce the land consumption of parking spaces and free up space for other uses [20]. In

2002 and 2008, vehicles in Germany were occupied by an average of 1.5 people [49]. One method of efficient use of passenger cars is the formation of car-pools and the operation of ridesharing companies. Needs-based use of various sorts of low emission vehicles can also serve to reduce fuel consumption. The latter measures would lead to an increase in energy and vehicle efficiency [48]. Another component in the future mobility mix could be Neighborhood Elec-tric Vehicles.

Numerous regulatory control measures are possible, for example congestion charges, aviation taxation and subsidies (such as a jet fuel tax and a departure tax), a reform of company car taxation, parking space management (for example thr-ough pay and display), or an extension of emissions trading to road traffic [48]. The introduction of speed limits, or lowering existing speed limits, would also have an impact on greenhouse gas emissions such as CO2 (carbon dioxide) and NOx (nitric oxide and nitrogen dioxide). Passenger cars consume a dispropor-tionately large amount of fuel at high speeds. A speed limit can also have secon-dary emissions-reducing effects, about which there is still considerable uncertainty: lower maximum speeds and longer travel times can contribute to a shift in traffic to rail and to the promotion of vehicles with lower engine power [50, 51]. The exter-nalities of traffic, namely the impact that air pollution caused by motor vehic-les has on society and the environment, must also be taken into account here [52].

The 2019 Dutch nitrogen emissions crisis, which indirectly caused the Dutch farmers' protests, convinced the government in November 2019 to lower the speed limits in the Netherlands on national roads to 100 kilometers per hour during the day, from 6 am to 7 pm. In the evening and at night the old speeds were maintained [53, 54]. Meanwhile, the State of the Netherlands v. Urgenda Foundation court case was decided in favour of its plaintiff Urgenda (initially in June 2015, upheld on appeal in October 2018, and finally confirmed by the Supreme Court of the Netherlands on 20 December 2019), who successfully forced the government to implement the necessary measures to reduce the Netherlands' CO2 emissions from 1990 levels by 25% by 2020. Although the government was free to choose which measures it would take to achieve this reduction, the plaintiff and other environ-mentalists had been suggesting throughout the legal process to lower the speed limit as one of several effective options to do so [55, 56]. Similar environmental arguments for speed limits have been proposed in Germany [57-59].

As one of several methods to mitigate the environmental impact of aviation, a shift to other modes of transport or a switch from short-haul air traffic to high-speed trains has been proposed [4]. In several countries in Europe, increasingly in the 2010s and early 2020s, some governments have even imposed a short-haul flight ban on all airlines, while many governmental agencies, commercial companies, universities, and NGOs have imposed restrictions or prohibitions on their employees to not take short-haul flights that can also be properly accomplished by train [60, 61].

In the field of urban planning, there are concepts for walkability, the compact city ('city of short distances'), New Urbanism (or its variant New Pedestrianism), and car-free living. In research policy, there are demands to give more consideration to the consequences of motorized private transport in the form of practice-oriented and solution-oriented research [62].

4.6. Further Development of Local Public Transport

According to a 2015 study by the Verkehrsclub Deutschland, local public transport in Germany was not customer-friendly enough. Cryptic route networks, opaque fare systems, ticket machines that cannot be operated, draft bus stops, and a lack of announcements about transfer and connection options was criticized. The club also called for better linking of local public transport with other modes of transportation. This included bike racks at bus stops, information on taking bikes on buses and trains, and options for switching to car-sharing providers. Furthermore, the synchronization of timetables was criticised, because it led to unnecessarily long waiting times for connecting buses or trains [63]. In 2012, several local public transport companies reportedly had been making efforts to improve the usability of ticket machines in Bavaria and Saxony [64, 65]. Against this background, Federal Transport Minister Alexander Dobrindt in 2017 called for electronic tickets and a uniform tariff system for all transport associations to be established by 2019 [66].

Since the 2010s, there have been frequent discussions on whether local public transport should be free of charge. The best-known example of free public transport is the Estonian capital Tallinn, where buses and trains have been free since 2013. By 2021, most counties in Estonia had also introduced free buses and trains [67]. Public transport is also free throughout Luxembourg. In Germany, the cities of Monheim am Rhein and Langenfeld, Rhineland were testing free public transport as of September 2021 [68].

Some cities have introduced mini electric buses, primarily in inner-city areas. The historic city centre of Aix-en-Provence, France is very narrow and closed to cars, taxis and normal bus traffic. In order to get people with restricted mobility to their destination, wheelchair-accessible electric minibuses are frequented there without a fixed timetable [69]. Likewise, in the medieval old town of Regensburg, only mini-buses are still driving around [70]. Furthermore, two self-propelled e-shuttles are in use in Regensburg's industrial park [71]. Berlin and Göppingen also want to supplement their local public transport with electric, highly automated minibuses [72, 73].

4.7. List of Aerial Tramways

In some cities, cable-ways are built as part of local public transit. Such cable-ways can be found in places such as:

- Medellín (Metrocable (Medellín)),
- La Paz (Mi Teleférico),
- New York (Roosevelt Island Tramway),
- Portland (Portland Aerial Tram),
- Algiers (Cableways of Algiers),
- Lisbon (see Funiculars in Lisbon),
- Brest (see Cableways of Brest),
- Bozen [74], London (Emirates Air Line (cable car)) and
- Ankara [75].

Cable cars are electrically operated and they have very low CO_2 emissions compared to other modes of transport. At 50% capacity, a cable car causes 27 grams of CO_2 per person and kilometer, a train with an electric locomotive 30 grams, a bus with a diesel engine 38.5 grams, and a car with a combustion engine even 248 grams. Furthermore, cable cars cause practically no noise pollution on the route, since the individual gondolas do not have their own drive, but are moved by a central motor housed in the station. In Germany, on the occasion of the Bundesgartenschau ('Federal Horticultural Show'), cable cars have emerged in Berlin (IGA Cable Car), Koblenz (see Koblenz cable car) and Cologne (Cologne Cable Car). Compared to underground or suburban trains, cable cars are relatively cheap and can be built quickly. As of November 2021, there are projects to build more cable cars to supplement local public transit in Berlin, Bonn, Düsseldorf, Cologne, Munich, Stuttgart and Wuppertal [76].

Continuous development is also affecting the rural areas as well. As a solution, what came into play were the integrated systems of public transport that is playing an important role in the development of rural areas, especially in post-communist countries [77].

4.8. Propulsion and Energy Transition in Transport

In order to achieve the energy transition in transport, it is considered necessary to refrain from burning petroleum-based fuel and to use more climate-friendly propulsion technologies or fuels [48]. Electricity from renewable sources, or e-fuels or biofuels produced from green electricity, can serve as substitutes for petrol and diesel fuel [4].

Since the overall efficiency of e-fuels is far lower than direct electrification via electric cars, the German Advisory Council on the Environment has recommended restricting the use of electricity-based synthetic fuels to air and shipping traffic in particular, in order not to increase electricity consumption too much. For example, hydrogen-powered fuel cell vehicles (FCVs) require more than twice as much energy per kilometer as battery electric vehicles (BEVs), and vehicles with combustion engines powered by power-to-liquid fuels even need between four and six times as much [78]. Battery vehicles therefore have significantly better energy efficiency than vehicles that are operated with e-fuels [79]. In general, electric cars consume around 12 to 15 kWh of electrical energy per 100 km, while conventionally powered cars use the equivalent of around 50 kWh per 100 km [80]. At the same time, the energy required for the production, transport and distribution of fuels such as petrol or diesel is also eliminated [81]. In China in particular, the switch from internal combustion engines to electro-mobility is being promoted for health reasons (to avoid smog) in order to counteract the massive air pollution in the cities [7].

According to Canzler & Wittowsky (2016), the propulsion transition could also become the central building block of Germany's Energiewende [6], While the switch to renewable energies is already underway worldwide, the energy transition in transport is proving more difficult, especially with the switch from oil to sustainable energy sources. However, disruptive technologies (such as the development of more powerful and cheaper batteries or innovations in the field of autonomous driving) and new business models (especially in the field of digitalization) can also lead to unpredictable, rapid and far-reaching changes in mobility [82]. New methods of getting around in urban traffic have also emerged:

Quadricycles combining muscle power and an electric motor

Light electric vehicles are small, light and drive electrically (Austria).

4.8.1. Vienna

Vienna, the capital of Austria, has been consistently developing into a city that is restructuring public space and promoting local public transport. Viennese urban planner Hermann Knoflacher has stated: 'The money comes on foot or by bike.' The economic use of space as parking spaces is inefficient. A car-free street increases the turn-over of restaurants, clothing stores and retailers. This would create new jobs [41].

Pedestrian zones, such as the Graben in Vienna (introduced in 1971), make a city attractive.

The attractiveness of public transport can be stimulated by lowering the price of an annual pass: in Vienna one can use public transport with a subscription fee of 1 euro a day [83, 84]. Between 2012 and 2018 the number of annual ticket holders increased from 373,000 to 780,000. At the same time as the changeover, the city began to invest more heavily in local transport. In July 2018, some German cities announced that they would follow the Viennese model and lower the prices for annual tickets [85].

The Mariahilfer Straße in Vienna in 2006, with frequent car congestion.

4.8.2. Luxembourg

Since 1 March 2020, local public transport across Luxembourg has been free of charge for everyone. The Grand Duchy thus became the first country in the world to introduce free local public transit [86]. An exception to this is first class tra-vel on the railways [87]. A major reason for the overhaul was the increasingly proble-matic traffic jams on Luxembourg's roads [88].

4.8.3. Further Examples

Cycling in Münster: a mobility station connects buses and bikes.

Several more significant examples of (potential) components and initiatives for mobility transition that have been proposed, studied, or put into practice include:

- As an alternative to the Viennese model of the annual ticket, a citizen ticket is being discussed in some German municipalities as a new way of financing and using local public transport. It is to be financed by a levy for all citizens of a municipality and function as a kind of flat rate for buses and trains [89].

- Phase-out of fossil fuel vehicles: In Germany, a ban on the sale of combustion engines from 2030 has been adopted by the Bundesrat in October 2016 [90]. Norway, on the other hand, already wants no cars with petrol or diesel engines to be registered from 2025 and ships and ferries only to be registered without fossil fuels from 2030, and is therefore considered a leading nation in electro-mobility. The Netherlands are also planning a ban on the registration of conventional drives in cars from 2025 [91]. In China, all automotive groups are obliged to meet a quota for the production and sale of purely electric or plug-in hybrid drives [92].
- There are numerous electro-mobility projects in Germany, such as the Modell regionen Elektro-mobilität and BeMobility. The German Association of Towns and Municipalities (DStGB) sees towns and municipalities as drivers and designers of the mobility transition and also supports a number of projects [93].

Critical Mass in Budapest (2007)

- Critical Mass is a form of direct action for promoting more and safer cycling in cities around the world. When riding together through inner cities, cyclists draw attention to cycling as a form of individual transport, advocate for mobility transition and, in particular, more rights for cyclists, better cycling traffic networks and infrastructure, and more room for non-motorized traffic. The first Critical Mass action took place in September 1992 in San Francisco [94].

- To improve air quality, efforts across Europe are being stepped up to introduce low-emission zones. A progressive approach is the French Crit'air, which provides for different restrictions depending on air pollution. The applicable prohibitions can be viewed on the Internet or via phone app. Electric vehicles or hydrogen-powered vehicles receive category 0 (green vignette) and can always drive anywhere [95]. Diesel driving bans were also issued in Germany.

- Instead of a company car, individual companies offer their employees a mobility budget that can be used to pay for different means of transport for business purposes.

- The city-state of Singapore has not allowed additional private cars since 1 February 2018 under its Vehicular Quota System. This is intended to promote the switch to other means of transport. It is the only country in the world which requires prospective vehicle owners to bid for a Certificate of Entitlement before they are allowed to own a vehicle for up to 10 years. The state only gives permission for a new car if another has been de-registered [96]. Singapore was also the first country in the world to implement congestion pricing in 1975 [97].

- Since 2003, there has been a London congestion charge which drivers have to pay in Central London. From October 2017 on, an additional, new fee for older and more polluting cars and vans is due with a toxicity charge [98, 99].

- In many cities in Germany there are citizens' initiatives which, following the example of the Initiative Volksentscheid Fahrrad ("Cycling Referendum Initiative") in Berlin, advocate for mobility transition and "bicycle laws" [100]. In June 2018, the Berlin Mobility Act to promote cycling was passed in Berlin, also due to a successful application for a referendum [101].

- Traffic lights are being tested in Karlsruhe as part of a pilot project which, in contrast to conventional pedestrian traffic lights, display a permanent green light for pedestrians and cyclists, not for vehicles, and only interrupt this when a vehicle approaches [102].

- In Japan, it is generally illegal to park a car on the street; a car buyer must provide evidence of owning private parking space or renting a public parking space for the car [103]. As of 2019, renting fees for public parking spaces in the more central districts of Tokyo cost about €300-500 a month, while in residential areas on the outskirts of Tokyo they cost around €100 a month [103]. Only after the police have verified that the parking lot exists and is large enough for the car the owner want to buy, the car dealer approves the purchase, and gives the owner a parking sticker to put on the new car's front or rear window [103]. The Japanese state has been using regulations to discourage the sale of luxury cars and to stimulate consumers to buy small light-weight cars with small engines or to motivate them to switch to local public transport [103].
- In Spain, a general speed limit of 30 km/h (19 mph) in built-up areas was introduced in 2021. On narrow streets with only one lane (often found in historic city centers), the permitted speed was limited to a maximum of 20 km/h (12 mph); for streets with more than one lane in both directions, the previously set speed limit was maintained at 50 km/h. A total of 509 people died in urban traffic accidents in Spain in 2019. The 2021 reduction of urban speed limits was intended to reduce the risk of pedestrians dying after being hit by a car by 80% [104].
- With the educational motto Weniger Wagen wagen ("risk fewer cars"), the Roman Catholic Archdiocese of Cologne has sought to raise awareness, and has calculated: 'Due to mobility (journeys to work, committees, church services, etc.), around 16,370 tons of CO_2 (as of 2012) are emitted annually in the Archdiocese of Cologne. This corresponds to a share of approx. 13 per cent of the archdiocese's total emissions.' In response, the Archdiocese stated it sought 'strategic and practical reorientation of mobility', including stimulating cycling through the Pharr-Rad initiative (a pun on Pfarrer "priest" and Fahrrad "bycicle") and the BistumsTicket ("diocesan ticket") which offers reduced fees for public transport travels by groups of 50 people or more to Catholic events organized within the archdiocese [105].

4.9. Short-haul Flight Ban

A short-haul flight ban is a prohibition imposed by governments on airlines to establish and maintain a flight connection over a certain distance, or by organizations or companies on their employees for business travel using existing flight connections over a certain distance, in order to mitigate the environmental impact of aviation (most notably to reduce anthropogenic greenhouse gas emissions which is the leading cause of climate change). In the 21st century, several governments, organizations and companies have imposed restrictions and even prohibitions on short-haul flights, stimulating or pressuring travelers to opt for more environ-mentally friendly means of transportation, especially trains [107].

The Belgian region of Wallonia banned Jet4you from flying between Charleroi and Liège in 2006 for environmental reasons [106].

A portion of air travelers in short-haul routes connect to other flights at their destination. A blanket ban would have a significant impact on these travelers [108], as inadequate rail connectivity between airports and main railway hubs of cities generally result in longer overall travel times and disruption to travellers overall [109].

By July 2019, most political parties in Germany, including the Left Party, the Social Democrats, the Green Party and the Christian Democrats, started to agree to move all governmental institutions remaining in Bonn (the former capital of West Germany) to Berlin (the official capital since German Reunification in 1990), because ministers and civil servants were flying between the two cities about 230,000 times a year, which was considered too impractical, expensive and environmentally damaging. The distance of 500 kilometers between Bonn and Berlin could only be travelled by train in 5.5 hours, so either the train connections required upgrading, or Bonn had to be abolished as the secondary capital [110, 111].

4.10. Measures in freight Transport

4.10.1. Sea Freight

By far the largest part of the world's freight traffic is sea freight. In 2010, about 60,000 trillion kilometer-tonnes were transported by sea, which was 85% of the world's total freight traffic. According to a 2015 forecast by Statista, by 2050 the volume of freight will have increased to four times the levels of 2010, while the share of sea freight will remain about the same [112].

Transporting goods by container ship is very efficient. Relatively few carbon dioxide (CO2) emissions are caused per transported tonne and kilometer compared the transport by truck (lorry). According to the Naturschutzbund Deutschland (NABU), the latter emit 50 grams of carbon dioxide per tonne and kilometer, while container ships only emit 15 grams [113]. However, the mineral oil-based ship fuel used by container ships is particularly polluting; 90 per cent of all large ships run on heavy fuel oil (bunker fuel). Among other things, this means that emissions of toxic sulfur oxide are many times higher [114]. To counteract this problem, the International Maritime Organization (IMO) lowered the limit value for sulfur in fuel from 3.5% to 0.5% in 2020 [115].

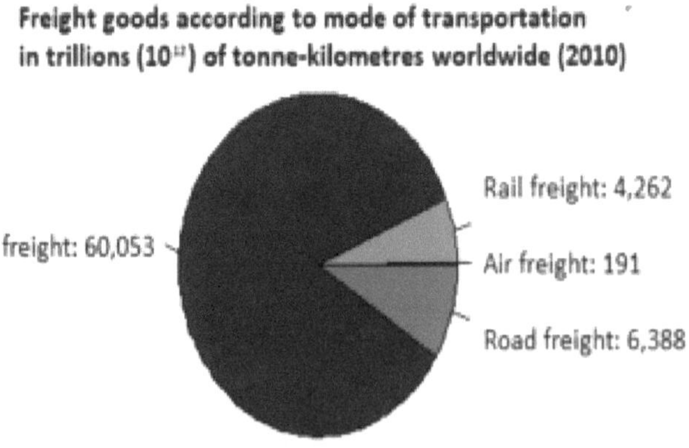

Global freight volumes according to mode of transport
in trillions of tonne-kilometres in 2010.

Efficiency can be further increased and fuel consumption reduced by building the ships even larger [114].

There are innovations to harness wind power for sea transportation. These include cylindrical sails that can be retrofitted to cargo ships (making them "rotor ships" or "Flettner ships") and can reduce fuel consumption. Another option is a towing kite construction, which was originally developed in 2001 by the Hamburg-based company SkySails and is now being sold by Air-Seas. The sail has an area of 1,000 square meters and was developed to reduce fuel consumption on cargo ships by up to 20%. As of 2019, the aviation group Airbus was testing this idea on four of its own freighters with the aim of saving up to 8,000 tonnes of carbon dioxide emissions [114].

4.10.2. Inland Navigation

As inland navigation (also known as 'inland waterway transport' (IWT) or 'inland shipping') is a relatively environmentally friendly option for freight transport (similar to rail freight transport), researchers and policy makers have been aiming to shift the volume of cargo transported by more pollutive means towards inland navigation (for example, as part of the 2019 European Green Deal) [116]. According to the Research Information System for Mobility and Traffic (FIS; an agency of the German Transport Ministry), deficits in the competitiveness of German inland navigation, especially in an international comparison, are responsible for the stagnating transport volume of German inland navigation. A water infrastructure that is not optimally developed with insufficient water channel depths and bridge clearance heights lead to low loading capacities and thus to high costs. A certain exception are the waterways of the Rhine area, which also have by far the highest transport volume [117]. Furthermore, the German inland waterway fleet is quite old by international comparison (45 years in 2013) [118].

Inland navigation is closely related to seaport hinterland traffic. For example, in the modal split in hinterland traffic at the Dutch and Belgian seaports (Rotterdam, Amsterdam, Antwerp and Zeebrugge), inland shipping has a share of around 55%, while in Germany it usually remains below 10% of hinterland traffic. The reason for this is the better expansion of the Rhine waterways. Furthermore, the majority of the 250 important inland ports in Germany are owned by large companies that only handle transport goods from third-party companies to a small extent [119]. Against this background, the FIS has called for the expansion and maintenance of German waterways. The number and carrying capacity of the German inland waterway vessels has remained constant in the early 21st century and was around 2.61 million tonnes in 2015 [118].

Various approaches to energy efficiency and air pollution reduction are being tested and researched in inland shipping. This includes propulsion configurations such as the father–son concept [120], diesel-electric hybrid drives [121], hydrodynamic optimizations [122], fuel water emulsion injection [123], SCR-catalysts, diesel particulate filters [124], gas-to-liquid fuels (GTL) [125] or Liquified Natural Gas (LNG) [126], some of which can also be used in combination and are suitable for retrofitting existing systems [127]. With an engine funding program, the German Transport Ministry supports inland navigation companies in the installation and retrofitting of low-emission engines or other emission-reducing technologies. The funding rate is up to 70% [128].

4.10.3. Road Freight and Modal Share

Freight in Germany (2019) (total: 700 tonne-kilometers) [130]:

- Truck (71.2%)
- Rail (19%)
- Inland ship (7.3%)
- Oil pipeline (2.5%)

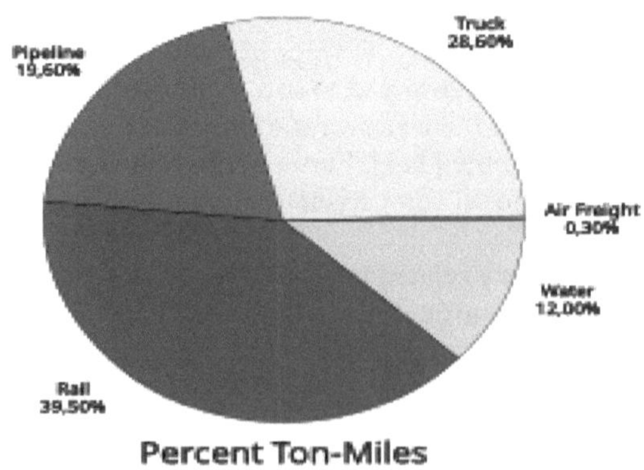

Freight in the United States by per cent ton-miles (2010 FRA report) [129].

In road freight transport, some transport companies are proposing partly new technologies such as trolley-trucks, electric trucks or electric cargo bikes. Package delivery services are experimenting with new concepts of smart logistics [131]. Trolley-trucks with an auxiliary battery offer the possibility of lower-emission long-distance truck transport that is also more energy-efficient than battery-powered trucks [132]. Equipping motorways with overhead lines for heavy goods vehicles (HGVs) has the advantage that HGVs would only have to carry small batteries, as only comparatively short distances would be covered in battery-only mode. At the same time, trolley-trucks would be a cost-effective way to make freight transport climate-friendly, as the electrification of motorways, at a cost of 3 million euros/km, does not represent too much of a financial outlay [133].

Another option to reduce CO2 emissions and environmental problems is to shift truck traffic to freight rail and inland waterway transport. This process is also known as modal shift. The German Environment Agency gives the climate impact of transport by truck in the reference year 2020 as 126 grams of CO2 equivalents per tonne-kilometer on average (g/tkm). According to the Environment Agency, transport by freight train has a climate impact of 33 g/tkm and transport by inland waterway vessel has a climate impact of 43 g/tkm, making rail and ship significantly more climate-friendly [134].

Although the European Union and its member states strongly promote the use of inland waterways and rail in combination with truck transport, in some cases financially [135], only HGVs have been developing positively in the 2010s, while shipping and rail have been stagnating or recording declines [136]. For 2016, the Federal Statistical Office of Germany reported a decline in transport performance of 3.7% for inland waterways, a decline of 0.5% for rail and growth of 2.8% for trucks [137]. In 2015, with a growing transport volume of 1.1%, there was a plus of 1.9% for road, a minus of 1% for rail and a minus of 3.2% for inland waterways [138]. Overall, 71% of the transport performance is accounted for by the truck [136].

With growing containerization however, a combination of different modes of transport (intermodal freight transport) becomes more efficient. In so-called multimodal transport or combined transport, the truck only has to cover the last mile between the port or rail terminal and the customer. Measures to promote combined transport are, for example:

- The Port of Rotterdam has set a quota for the modal share of hinterland transport modes: the truck share is to drop from 47% to 35%, while rail is to provide 20% instead of 13% in the future, and the transport performance of inland waterways is to increase from 40% to 45% [139].
- Instead of burdening trunk roads with the transport of heavy goods such as industrial plants or components for wind turbines, German transport companies have been required since 2010 to use the electronic portal Procedural Management of Large and Heavy Goods Transport (VEMAGS) to check whether alternative transport routes such as ship and rail are available, and if not, to explain that in their application for a permit to transport goods via road trucks [140].
- With the promotion of handling facilities for combined transport, the German federal government supports the shift in traffic to inland waterways and freight trains [135].
- The Lower-Rhine Chamber of Commerce and Industry, the Schifferbörse and the Development Centre for Naval Technology and Transport Systems (DST) in Duisburg jointly offer an additional course. Apprentice forwarding and logistics clerks should thus learn about the advantages of alternative modes of transport, rail and inland waterway, and thus integrate them more easily into their everyday work. Frequently, the curriculum only includes road freight transport and additional sea freight or air transport [141].

4.11. References

[1]. Adey et al. 2021, p. 4.
[2]. Shiftan, Yoram (2016). Transition towards Sustainable Mobility : Role of Instruments, Individuals and Institutions. Abingdon: Routledge. pp. 65–68. ISBN 9781317007302. Retrieved 26 February 2022.
[3]. Gianna Niewel (2 March 2020). "Verkehrswende in München – Das Drama einer Umverteilung". Süddeutsche Zeitung. Retrieved 5 March 2020.
[4]. Johan Rockström (2017), "A roadmap for rapid decarbonization". Science, vol. 355, no. 6331, pp. 1269–1271.
[5]. Henrik Mortsiefer (9 May 2017). ""Es wird ernst"". Der Tagesspiegel. Retrieved 22 September 2020.
[6]. Weert Canzler, Dirk Wittowsky (2016), "The impact of Germany's Energiewende on the transport sector – Unsolved problems and conflicts", Utilities Policy (in German), vol. 41, pp. 246–251.
[7]. Manfred Kriener (9 February 2017). "China elektrisiert. Ein Blick in die Zukunft Autoindustrie". Le Monde diplomatique. Retrieved 13 April 2017.
[8]. Claudia Hornberg et al. (2017), Sachverständigenrat für Umweltfragen [SRU] (ed.), Umsteuern erforderlich: Klimaschutz im Verkehrssektor: Sondergutachten November 2017. Berlin, ISBN 978-3-947370-11-5. Retrieved 20 September 2020. Retrieved 22 September 2020.
[9]. Johannes Lelieveld (2015), "The contribution of outdoor air pollution sources to premature mortality on a global scale", Nature, vol. 525, no. 7569, pp. 367–371.
[10]. Johannes Lelieveld (2017), "Clean air in the Anthropocene", Faraday Discussions, vol. 200, pp. 693–703.
[11]. Bundesamt, Statistisches (2022). "Road Transport: Car Dominance Unbroken". destatis.de. Federal Statistical Office of Germany. Retrieved 21 October 2022.
[12]. Harvey, Fiona (5 March 2020). "One in five Europeans exposed to harmful noise pollution – study". The Guardian. Retrieved 22 September 2020.
[13]. ""Nachhaltigen Verkehr fördern – Verkehrssysteme der Zukunft entwickeln – Investitionsoffensive von Bund und Ländern notwendig"". Deutscher ' Städtetag. 22 June 2018. Retrieved 22 September 2020.
[14]. "Verkehrswende in Stadt und Umland" (in German). Deutsches Institut für Urbanistik (Difu). June 2017. Archived on 7 October 2017. Retrieved 9 May 2017.
[15]. Günther, Regine; Krane, Volker (10 July 2020). "Verkehrswende: "Es geht nicht um Autofahrer gegen Radfahrer"". Die Zeit. Interviewed by Götz,

Sören. Retrieved 10 September 2023.
[16]. Higher social class predicts increased unethical behavior, Piff et al., Proceedings of the National Academy of Sciences Mar 2012, 109 (11) 4086-4091.
[17]. "Studie: Aggressivität im Verkehr steigt leicht". RedaktionsNetzwerk Deutschland. 1 October 2020. Retrieved 10 September 2023.
[18]. Sorge, Nils-Viktor (13 September 2018). "Aggressives Auto-Design: "Manche Autos wirken wie eine geladene Waffe"". Der Spiegel. ISSN 2195-1349. Retrieved 10 September 2023.
[19]. von Schneidemesser, Dr. Dirk (22 April 2021). "Wir brauchen eine neue Sprache für die Verkehrsberichterstattung". RIFS Potsdam. Retrieved 10 September 2023.
[20]. Peter Ilg (26 September 2016). "30 Prozent der Autos würden reichen". Die Zeit. Retrieved 22 September 2020.
[21]. Jean-Paul Rodrigue (2020), "Chapter 3 – Transportation, Economy and Society > Transportation and Society > Passengers Mobility Transition". The Geography of Transport Systems (5 ed.), New York: Routledge, (456 pages), ISBN 978-0-367-36463-2, retrieved 20 September 2020.
[22]. Adey et al. 2021, pp. 2–4.
[23]. Cox, Peter; Koglin, Till (2020). The Politics of Cycling Infrastructure:] Spaces and (In)Equality. Bristol: Policy Press. pp. 122–126. ISBN 9781447345183. Retrieved 28 February 2022.
[24]. Kenworthy, Jeffrey R. (2010). "Box. 8.7 Kenneth R. Schneider: Fighting for change". An Introduction to Sustainable Transportation: Policy, Planning Implementation. London/Washington, D.C.: Earthscan/Routledge. p. 254. ISBN 9781136541940. Retrieved 27 February 2022.
[25]. Zelinsky, Wilbur (April 1971). "The Hypothesis of Mobility Transition". Geographical Review. 61 (2): 219–249.
[26]. Adey et al. 2021, pp. 16–19.
[27]. "Vom "Gehzeug" zum "Fahrzeug"". Wiener Zeitung. 15 October 2014. Retrieved 22 September 2020.
[28]. "Buchtipp: "Virus Auto" von Hermann Knoflacher". mobilservice.ch. 6 April 2010. Retrieved 8 May 2017.
[29]. "Verkehrswende, die". Duden. Retrieved 22 September 2020.
[30]. Stiftung Mercator ("Mercator Foundation", named after Gerardus Mercator) is a German non-profit organisation which promotes equal opportunities, social cohesion, respect, tolerance, cosmopolitanism and protection of nature and the environment.
[31]. Christian Hochfeld, et al. (2017). Mit der Verkehrswende die Mobilität von

morgen sichern. 12 Thesen Verkehrswende. Berlin: Agora Verkehrswende. Retrieved 20 September 2020.
[32]. Mark Siemons (12 June 2016). "Bald eine Fahrradstadt? Berlin Kamikaze". Frankfurter Allgemeine Zeitung (in German). Retrieved 22 September 2020.
[33]. Katharina Manderscheid (2020), Achim Brunnengräber, Tobias Haas (ed.), "Antriebs-, Verkehrs- oder Mobilitätswende? Zur Elektrifizierung des Automobilitätsdispositivs", Baustelle Elektromobilität – Sozialwis-senscha-ftliche Perspektiven auf die Transformation der (Auto-) Mobilität, Bielefeld: transcript, pp. 37-67, ISBN 978-3-8376-5165-2, Retrieved 11 August 2020.
[34]. Gerald Traufetter, Florian Gathmann (26 April 2018). "Verkehrsminister Scheuer im Interview: "Verbote sind für mich kein Politikstil"". Der Spiegel. Retrieved 10 August 2020.
[35]. Andor, Dr. Mark (11 August 2020). "Warum wissen wir nicht, wie viel ein Auto im Monat kostet?". AufRuhr Magazin (Interview). Interviewed by Beck, Maren. Stiftung Mercator. Retrieved 10 September 2023.
[36]. Canzler, Weert; Knie, Andreas (February 2014). "Die Energie- und Verkehrswende in "Schlauen Netzen"". Innovationszentrum für Mobilität und gesellschaftlichen Wandel. Archived on 23 April 2017. Retrieved 23 April 2017.
[37]. Umweltbundesamt, ed. (2016), Klimaschutzbeitrag des Verkehrs bis 2050, Texte 56/2016 Umweltbundesamt, Dessau-Roßlau, ISSN 1862-4804. Retrieved 22 September 2020.
[38]. "Bundesverkehrswegeplan: "Es ist keine Verkehrswende"". Deutschlandfunk. 3 August 2016. Retrieved 22 September 2020.
[39]. "Experte: "Verkehrspolitik auf dem Holzweg"". heute.de. Retrieved 21 April 2017.
[40]. Martin Stallmann (27 October 2021). "Umweltschädliche Subventionen: fast die Hälfte für Straßen- und Flugverkehr". Das Umweltbundesamt. Retrieved 31 October 2021.
[41]. "Verkehrswende: Wie sich die Blechlawine stoppen lässt". FAZ.NET. ISSN 0174-4909. Retrieved 6 January 2023.
[42]. "Greenpeace-Studie: Homeoffice kann Millionen Tonnen CO2 einsparen". Frankfurter Allgemeine Zeitung. 20 Aug. 2020. Retrieved 22 Sept. 2020.
[43]. "Ist die Coronakrise gut für die Verkehrswende?". Bundesministerium für Bildung und Forschung. 12 May 2020. Retrieved 22 September 2020.
[44]. "Beschleunigt Corona die Verkehrswende?". WDR. 19 May 2020. Retrieved 22 September 2020.
[45]. "Wie schafft Deutschland die Verkehrswende?". Deutsche Welle. 28 March 2017. Retrieved 23 September 2020.

[46]. Christian Frahm, Emil Nefzger (5 November 2019). "So viel kostet Autofahren - auch Menschen, die nicht Auto fahren". Spiegel Online. Retrieved 7 November 2019.
[47]. Huber, B. (2022). "Die Bahn bleibt das Rückgrat der Verkehrswende". System Bahn. Retrieved 10 September 2023.
[48]. "Metaanalyse über Maßnahmen Instrumente Energiewende im Verkehr". forschungsradar.de. Agentur für erneuerbare Energien (AEE). July 2016. Archived on 16 November 2019. Retrieved 13 April 2017.
[49]. German Aerospace Center (DLR), Institut für angewandte Sozialwissensh-aft (infas), ed. (February 2010), Mobilität in Deutschland 2008 (MiD2008): Ergebnisbericht. Struktur–Aufkommen–Emissionen–Trends. Bonn/Berlin. Retrieved 22 September 2020.
[50]. "Energiegipfel im Kanzleramt". Verkehrsclub Deutschland. 9 October 2006. Retrieved 17 May 2017.
[51]. Gunnar Gohlisch, Marion Marlow (June 1999), Umweltbundesamt (ed.), Umweltauswirkungen von Geschwindigkeitsbeschränkungen. Texte 40/1999 Umweltbundesamt. Berlin, pp. 23-29, ISSN 1862-4804. Retrieved 22 September 2020.
[52]. Reidl, Andrea (26 January 2022). "Soziale Kosten des Straßenverkehrs: Was Autofahren wirklich kostet". Die Zeit. Retrieved 10 September 2016.
[53]. Bastiaan Nagtegaal (13 November 2019). "Stikstofaanpak: landbouwinnovatie, noodwet voor de bouw en 'rotmaatregel' snelheid". NRC Handelsblad. Retrieved 2 March 2022.
[54]. Frommeyer, Lena (14 November 2019). "Tempolimit in den Niederlanden: Bremsen für den Fortschritt" [Speed limit in the Netherlands]. Der Spiege. Retrieved 2 March 2022.
[55]. SPIEGEL, Claus Hecking, DER (20 December 2019). "Niederlande: Klimaaktivisten siegen gegen Regierung - Der Spiegel - Wissenschaft". Der Spiegel. Retrieved 26 January 2020.
[56]. "Klimaklage in den Niederlande - Ein historisches Urteil". Deutschlandfunk. Retrieved 26 January 2020.
[57]. "Stauforscher, Klimaschützer und Unfallforscher zum Tempolimit sagen". app.handelsblatt.com. Retrieved 22 May 2021.
[58]. "Das wissen wir über Tempolimits". quarks.de. 16 January 2020. Retrieved 22 May 2021.
[59]. Götz, Sören (17 July 2020). "Weniger CO2 und weniger Unfälle". Die Zeit. Retrieved 22 May 2021.
[60]. Matthias Wabl and Christopher Jasper (9 June 2020). "Airline bailouts point to greener travel—and higher fares". BNN Bloomberg. Retrieved 13

June 2020.
[61]. "Exemplary measures taken by universities and public administrations". Unter 1000. Scientists 4 Future. Retrieved 26 October 2020.
[62]. "Mobil bleiben: Forschung bereit für Anwendung". Bundesministerium für Bildung und Forschung (BMBF). 22 June 2017. Retrieved 22 September 2020.
[63]. "Studie: Öffentlicher Nahverkehr zu wenig kundenfreundlich". evangelisch.de. 4 November 2015. Retrieved 2 March 2022.
[64]. "Ein Klick für Nürnberg". August 2012. Retrieved 19 November 2021.
[65]. "i-com: Nutzerzentrierte Entwicklung des Bediendialoges für die Münchner Fahrkartenautomaten". Nürnberger Nachrichten / Quantic.de. August 2012. Retrieved 2 March 2022.
[66]. "Öffentlicher Nahverkehr Mehr Transparenz und digitale Tickets". Deutschlandfunk Kultur. 6 January 2017. Retrieved 2 March 2022.
[67]. Kristin Becker and Marcel Heberlein (9 September 2021). "Mobilität und Klimaschutz: Wenn der Nahverkehr kostenlos wäre?". Tagesschau. Retrieved 2 March 2022.
[68]. "In diesen Städten sind Bus und Bahn kostenlos". 2 March 2020. Retrieved 19 November 2021.
[69]. "Diablines: Die Innenstadt-Busse von Aix". 31 October 2021. Retrieved 19 November 2021.
[70]. "E-Bus "emil" für die Regensburger Altstadt". Retrieved 19 Nov. 2021.
[71]. "Das ist der Startplan für autonome Busse". Retrieved 19 November 2021.
[72]. "Die "kleinen Gelben" auf neuem Kurs in Alt-Tegel". 29 June 2021. Retrieved 19 November 2021.
[73]. ""e-Bürgerbus": Bürgerschaftliches Engage kombiniert Elektromobilität". Retrieved 19 November 2021.
"Zwei Seilbahnen in der Landeshauptstadt". Retrieved 19 November 2021.
[74]. "Per Seilbahn durch Ankara". 26 March 2014. Retrieved 19 Nov. 2021.
[75]. Magazin für Energieversorgung - Planung - Bau - Betrieb - Service (ed.), Netzpraxis November 2021 - Mit gutem Gewissen über den Stau hinweg schweben, pp. 74ff
[76]. Šťastná, Milada; Vaishar, Antonín (1 September 2017). "The relationship between public transport and the progressive development of rural areas". Land Use Policy. 67: 107–114. ISSN 0264-8377.
[77]. Claudia Hornberg et al. (2017), Sachverständigenrat für Umweltfragen. Umsteuern erforderlich: Klimaschutz im Verkehrssektor: Sondergutachten November 2017. Berlin, ISBN 978-3-947370-11-5, retrieved 20 September 2020, "Umsteuern erforderlich: Klimaschutz im

Verkehrssektor". 23 November 2017. Retrieved 22 September 2020.
[78]. Volker Quaschning (2016), Hochschule für Technik und Wirtschaft Berlin (ed.), Sektorkopplung durch die Energiewende. Anforderungen an den Ausbau erneuerbarer Energien zum Erreichen der Pariser Klimaschutz-ziele unter Berücksichtigung der Sektorkopplung. Retrieved 22 Sept. 2020.
[79]. Günther Brauner (2016), Energiesysteme: regenerativ und dezentral. Strategien für die Energiewende, Wiesbaden: Springer Vieweg, p. 119. ISBN 978-3-658-12754-1
[80]. Andreas B. (2019). "Endenergiebezogene Analyse Diesel versus Elektromo-bilität". Springer-Professionals. Retrieved 22 September 2020.
[81]. "VW: "The next big thing"". 3sat 7 October 2016. Retrieved 22 Sept. 2020.
[82]. Peter Münch (11 February 2018). "Die Wiener "Öffis" kosten nur einen Euro am Tag". Süddeutsche Zeitung. Retrieved 22 September 2020.
[83]. "Jahreskarte". Wiener Linien. Retrieved 11 February 2018.
[84]. Melanie Berger (30 June 2018). "Öffentlicher Nahverkehr – Jahresticket 365 Euro: Deutsche Städte mögen Wiener Modell". Der Tagesspiegel. Retrieved 22 September 2020.
[85]. Kirsch, Matthias (29 February 2020). "Gratis-ÖPNV in Luxemburg: Es hapert mit dem Vorzeigeprojekt". Der Spiegel. ISSN 2195-1349. Retrieved 6 January 2023.
[86]. sda/mr: Öffentlicher Verkehr in Luxemburg wird kostenlos. In: Eisenbahn-Revue International 3/2019, S. 153.
[87]. NACHRICHTEN, n-tv. "Kostenloser ÖPNV allein bringt nichts". n-tv.de. Retrieved 6 January 2023.
[88]. René Rabenschlag (2016). "Bürgerticket: Flatrate für Bus und Bahn?". Westdeutscher Rundfunk. Archived on 26 May 2017. Retrieved 21 April 2017.
[89]. Sven B. (2016). "Bundesländer wollen Benzin- und Dieselautos verbieten". Der Spiegel. Archived on 26 September 2020. Retrieved 19 Sept. 2020.
[90]. "Norwegen will Benzin- und Dieselautos verbieten". Spiegel Online. 16 August 2016. Retrieved 23 September 2020.
[91]. Steffen Wurze (2017). "Weicht China die Quote für E-Autos auf?". Tagesschau.de. Retrieved 23 April 2017.
[92]. "Neue Mobilität – Kommunen als Akteure der Verkehrswende". Deutscher Städte- und Gemeindebund. 17 February 2017. Retrieved 4 May 2017.
[93]. Garofoli, Joe (September 28, 2002). "Critical Mass turns 10". San Francisco Chronicle. Archived on January 20, 2008. Retrieved July 2, 2007.
[94]. "Umweltzonen: Durchblick mit der Green-Zones-App". KFZ-Anzeiger. 8 May 2017. Archived on 16 February 2018. Retrieved 9 May 2017.

[95]. "Govt adopts zero-growth stance for car, motorcycle populations". The Straits Times. Archived on 25 May 2024. Retrieved 13 January 2025.
[96]. "Distance-based pricing: Best solution to traffic congestion?". AsiaOne. 12 November 2023. Retrieved 13 January 2025. Singapore was actually the first country in the world to implement congestion charges, back in 1975.
[97]. Björn Finke (2018). "Die Giftmaut soll Londons Verkehrsprobleme lösen". Süddeutsche Zeitung. Retrieved 22 September 2020.
[98]. Kimiko de Freytas-Tamura (17 February 2017). "A Push for Diesel Leaves London Gasping Amid Record Pollution". New York Times. Retrieved 22 September 2020.
[99]. Stefan Tomik (2018). "Verkehrswende-Initiative: Mehr Fahrrad wagen". Frankfurter Allgemeine Zeitung. Retrieved 22 September 2020.
[100]. Jörn Hasselmann, Klaus Kurpjuweit (27 June 2018). "Mobilitätsgesetz: Berlin regelt den Verkehr neu". Der Tagesspiegel. Retrieved 19 Aug. 2018.
[101]. "Schritt zur Verkehrswende: Immer grün für Fußgänger – Pilotprojekt in Karlsruhe". heise.de. 1 January 2022. Retrieved 7 January 2022.
[102]. Martin Kölling (6 March 2019). "Autos, die wie Toastbrote aussehen – Japans Autopolitik Trump auf". Handelsblatt. Retrieved 28 March 2022.
[103]. dpa (2021). "Spanien führt 30-km/h-Tempolimit innerorts ein". heise.de. Heinz Heise. Retrieved 28 March 2022.
[104]. "Emissionsarm mobil sein". erzbistum-koeln.de. Archdiocese of Cologne. September 2017. Retrieved 28 March 2022.
[105]. "Jet4You mag niet van Charleroi naar Luik vliegen". De Morgen. 28 Nov. 2008. Retrieved 21 October 2020.
[106]. Wabl, Matthias; Jasper, Christopher (9 June 2020). "Airline bailouts point to greener travel—and higher fares". BNN Bloomberg. Retrieved 13 June 2020.
[107]. Jon Sindreu (2022). "Why Trains Will Struggle to Replace Short Flights". The Wall Street Journal. Retrieved 14 October 2022.
[108]. Reiter, Vreni; Voltes-Dorta, Augusto; Suau-Sanchez, Pere (2022). "The substitution of short-haul flights with rail services in German air travel markets: A quantitative analysis". Case Studies on Transport Policy. 10 (4): 2025–2043.
[109]. Jorg Luyken (23 July 2019). "Calls to end Bonn's status as Germany's second city as Greens seek to ban all domestic flights". The Daily Telegraph. Retrieved 23 October 2020.
[110]. "German government employees increased air travel in 2019". Deutsche Welle. 2 August 2020. Retrieved 23 October 2020.
[111]. "Frachtgut nach Verkehrsträger weltweit". Retrieved 11 September 2021.

[112]. "Mythos klimafreundliche Containerschiffe". Naturschutzbund Deutschland. Retrieved 11 September 2021.
[113]. Thorge Jans (6 March 2019). "Nachhaltige Konzepte für Containerschiffe". Retrieved 11 September 2021.
[114]. "Neuer Grenzwerkt Seeschiffkraftstoffe". Umweltbundesamt. 15 Jan. 2020. Retrieved 11 September 2021.
[115]. Hofbauer, Florian; Putz, Lisa-Maria (2020). "External Costs in Inland

Waterway Transport: An Analysis of External Cost Categories and Calculation Methods". Sustainability. 12 (5874). MDPI: 9. Retrieved 29 March 2022.
[116]. "Gebietsstruktur der Binnenschifffahrt und deutsche Binnenhäfen". Retrieved 12 September 2021.
[117]. "Gründe für die Ertüchtigung der Binnenschifffahrt". Retrieved 12 Sept. 2021.
[118]. "Gründe für die Instandhaltung und den Ausbau der deutschen Binnenwasserstraßen". Retrieved 12 September 2021.
[119]. Christian Grohmann (29 September 2015). "Rolf Bach erhält Innovationspreis der Allianz Esa". bonapart. Retrieved 23 September 2020.
[120]. Christian G. (2013). "Dieselelektrisches Gütermotorschiff "Goblin" Fahrt". bonapart. Retrieved 23 September 2020.
[121]. Christian Grohmann (15 August 2015). ""Rhenus Duisburg" fährt mit Flex-Tunnel und vier Motoren". bonapart. Retrieved 23 September 2020.
[122]. Christian Grohmann (2016). "Kraftstoff-Wasser-Emulsionstechnik macht "Rudolf Deymann" zum schwimmenden Exponat". bonapart. Retrieved 23 September 2020.
[123]. Christian Grohmann (6 November 2015). "Laborschiff "Max Prüss" erhält Abgasreinigungsanlage". bonapart. Retrieved 23 September 2020.
[124]. Christian Grohmann (26 April 2016). "Emissionsarm ohne Umbau: "Jenny" bunkert synthetischen GTL-Kraftstoff". bonapart. Retrieved 23 Sept. 2020.
[125]. Christian Grohmann (15 November 2013). "Erste LNG-Bunkerung in Deutschland im Hafen Mannheim durchgeführt". bonapart. Retrieved 23 September 2020.
[126]. Christian Grohmann (20 May 2016). "Deutz 545 fährt schadstoffarm mit Kraftstoff-Wasser-Emulsion". bonapart. Retrieved 23 September 2020.
[127]. Christian Grohmann (4 August 2015). "Motorenförderprogramm: Jetzt bis zu 70 Prozent vom Staat". bonapart. Retrieved 23 September 2020.
[128]. "National Rail Plan Progress Report | FRA". railroads.dot.gov. Sept. 2010.

Retrieved 31 March 2022.
[129]. "Anteile der Verkehrsträger am Güterverkehr; 2019 in Deutschland". Allianz pro Schiene. November 2020. Retrieved 12 September 2021.
[130]. "Hamburg sucht nach intelligenteren Wegen für Paketzustellung". VerkehrsRundschau. 25 November 2016. Retrieved 22 September 2020.
[131]. Jens Tartler (11 April 2017). "Hendricks fährt auf Strom-LKW ab". Der Tagesspiegel. Retrieved 22 September 2020.
[132]. Sonja Fröhlich (7 May 2019). "Erste Elektro-Autobahn für Lastwagen – eine ökologische Alternative?". Hannoversche Allgemeine Zeitung. Retrieved 22 September 2020.
[133]. "Umweltfreudlich mobil! Ein ökologischer Verkehrsartenvergleich für den Personen- und Güterverkehr in Deutschland". Umweltbundesamt. March 2020. Retrieved 11 September 2021.
[134]. Christian Grohmann (9 January 2017). "KV-Förderrichtlinie 2017–2021 in Kraft getreten". bonapart. Retrieved 22 September 2020.
[135]. "Beförderungsmenge und Beförderungsleistung nach Verkehrsträgern". Statistisches Bundesamt (Destatis). 19 May 2020. Retrieved 22 Sept. 2020.
[136]. "Güterverkehr 2016: Neuer Höchststand beim Transportaufkommen". Statistisches Bundesamt (Destatis). 17 Feb. 2017. Retrieved 19 May 2017.
[137]. Christian Grohmann (19 February 2016). "Güterverkehr 2015 wächst – Binnenschiff und Bahn verlieren". bonapart. Retrieved 19 May 2017.
[138]. "Schiene spielt Schlüsselrolle für Rotterdam". Deutsche Verkehrs-Zeitung (DVZ). 2 March 2015. Retrieved 19 May 2017.
[139]. Christian Grohmann (13 February 2017). "Schwergut und Projektladungen: Leichtes Wachstum in 2017 möglich". bonapart. Retrieved 23 Sept. 2020.
[140]. Axel Götze-Rohen (25 February 2015). "Schifferbörse startet Qualifikationsinitiative". bonapart. Retrieved 23

Chapter (5)
Hydrogen Economy

5.1. Preface

Hydrogen has the most potential to reduce greenhouse gas emissions when used in chemical production, refineries, international shipping, and steelmaking [1].

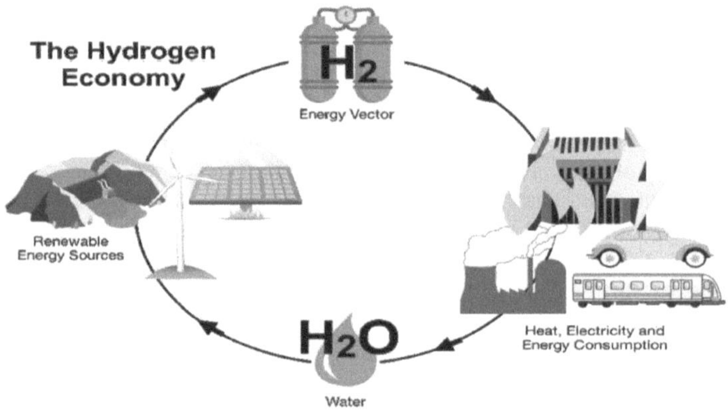

The hydrogen economy is an umbrella term for the roles hydrogen can play alongside low-carbon electricity to reduce emissions of greenhouse gases. The aim is to reduce emissions where cheaper and more energy-efficient clean solutions are not available [2]. In this context, hydrogen economy encompasses the production of hydrogen and the use of hydrogen in ways that contribute to phasing-out fossil fuels and limiting climate change.

Hydrogen can be produced by several means. Most hydrogen produced today is gray hydrogen, made from natural gas through steam methane reforming (SMR). This process accounted for 1.8% of global greenhouse gas emissions in 2021 [3]. Low-carbon hydrogen, which is made using SMR with carbon capture and storage (blue hydrogen), or through electrolysis of water using renewable power (green hydrogen), accounted for less than 1% of production [4]. Virtually all of the 100 million tonnes [5] of hydrogen produced each year is used in oil refining (43%

in 2021) and industry (57%), principally in the manufacture of ammonia for fertilizers, and methanol [6].

To limit global warming, it is generally envisaged that the future hydrogen economy replaces gray hydrogen with low-carbon hydrogen. As of 2024 it is unclear when enough low-carbon hydrogen could be produced to phase-out all the gray hydrogen [7]. The future end-uses are likely in heavy industry (e.g. high-temperature processes alongside electricity, feedstock for production of green ammonia and organic chemicals, as alternative to coal-derived coke for steel-making), long-haul transport (e.g. shipping, and to a lesser extent hydrogen-powered aircraft and heavy goods vehicles), and long-term energy storage [8, 9]. Other applications, such as light duty vehicles and heating in buildings, are no longer part of the future hydrogen economy, primarily for economic and environmental reasons [10, 11]. Hydrogen is challenging to store, to transport in pipelines, and to use. It presents safety concerns since it is highly explosive, and it is inefficient compared to direct use of electricity. Since relatively small amounts of low-carbon hydrogen are available, climate benefits can be maximized by using it in harder-to-decarbonize applications [11].

As of 2023 there are no real alternatives to hydrogen for several chemical processes in which it is currently used, such as ammonia production for fertilizer [12]. The cost of low- and zero-carbon hydrogen is likely to influence the degree to which it will be used in chemical feed-stocks, long haul aviation and shipping, and long-term energy storage. Production costs of low- and zero-carbon hydrogen are evolving. Future costs may be influenced by carbon taxes, the geography and geopolitics of energy, energy prices, technology choices, and their raw material requirements. It is likely that green hydrogen will see the greatest reductions in production cost over time [13]. The U.S. Department of Energy's Hydrogen Hot-shot Initiative seeks to reduce the cost of green hydrogen drop to $1 a kilogram during the 2030s [14].

5.2. History and Objectives

5.2.1. Origins

The concept of a society that uses hydrogen as the primary means of energy storage was theorized by geneticist J. B. S. Haldane in 1923. Anticipating the exhaustion of Britain's coal reserves for power generation, Haldane proposed a network of wind turbines to produce hydrogen and oxygen for long-term energy storage through electrolysis, to help address renewable power's variable output

[15]. The term "hydrogen economy" it was coined by John Bockris during a talk he gave in 1970 at General Motors (GM) Technical Center [16]. Bockris viewed it as an economy in which hydrogen, underpinned by nuclear and solar power, would help address growing concern about fossil fuel depletion and environmental pollution, by serving as energy carrier for end-uses in which electrification was not suitable [17].

A hydrogen economy was proposed by the University of Michigan to solve some of the negative effects of using hydrocarbon fuels where the carbon is released to the atmosphere (as carbon dioxide, carbon monoxide, unburnt hydrocarbons, etc.). Modern interest in the hydrogen economy can generally be traced to a 1970 technical report by Lawrence W. Jones of the University of Michigan [18], in which he echoed Bockris' dual rationale of addressing energy security and environmental challenges. Unlike Haldane and Bockris, Jones only focused on nuclear power as the energy source for electrolysis, and principally on the use of hydrogen in transport, where he regarded aviation and heavy goods transport as the top priorities [19].

5.2.2. Later Evolution

A spike in attention for the hydrogen economy concept during the 2000s was repeatedly described as hype by some critics and proponents of alternative technologies [21-23], and investors lost money in the bubble [24]. Interest in the energy carrier resurged in the 2010s, notably with the forming of the World Hydrogen Council in 2017. Several manufacturers released hydrogen fuel cell cars commercially, with manufacturers such as Toyota, Hyundai, and industry groups in China having planned to increase numbers of the cars into the hundreds of thousands over the next decade [25, 26].

The global scope for hydrogen's role in cars is shrinking relative to earlier expectations [27, 28]. By the end of 2022, 70,200 hydrogen vehicles had been sold worldwide [29], compared with 26 million plug-in electric vehicles [30].

Early 2020s takes on the hydrogen economy share earlier perspectives' emphasis on the complementarity of electricity and hydrogen, and the use of 5.4. electrolysis as the mainstay of hydrogen production. They focus on the need to limit global warming to 1.5 °C and prioritize the production, transportation and use of green hydrogen for heavy industry (e.g. high-temperature processes alongside electricity [31], feedstock for production of green ammonia and organic chemicals, as alternative to coal-derived coke for steelmaking) [32], long-haul transport (e.g. shipping, aviation and to a lesser extent heavy goods vehicles), and long-term energy storage.

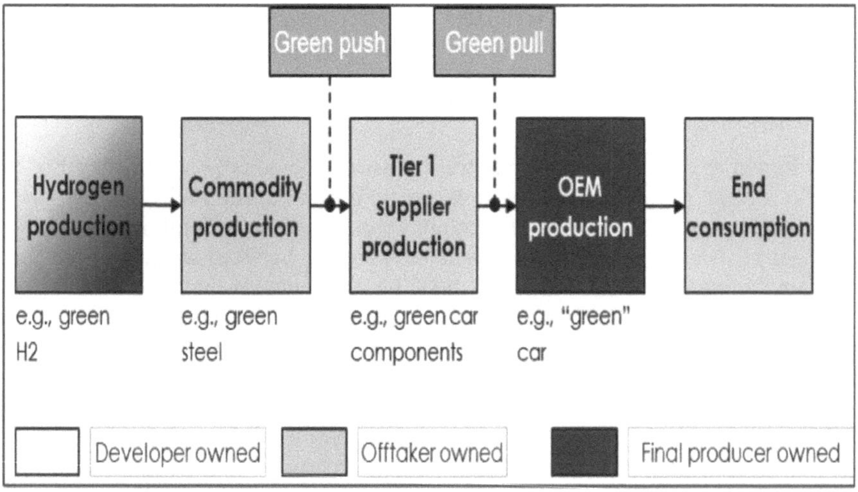

Technology leadership opportunities in green hydrogen value chains according to the International Renewable Energy Agency in 2022 [20].

5.3. Current Hydrogen Market

Hydrogen production globally was valued at over US$155 billion in 2022 and is expected to grow over 9% annually through 2030 [33].

In 2021, 94 million tonnes (Mt) of molecular hydrogen (H2) was produced [34]. Of this total, approximately one sixth was as a by-product of petrochemical industry processes. Most hydrogen comes from dedicated production facilities, over 99% of which is from fossil fuels, mainly via steam reforming of natural gas (70%) and coal gasification (30%, almost all of which in China). Less than 1% of dedicated hydrogen production is low carbon: steam fossil fuel reforming with carbon capture and storage, green hydrogen produced using electrolysis, and hydrogen produced from biomass. CO2 emissions from 2021 production, at 915 MtCO2 [35], amounted to 2.5% of energy-related CO2 emissions [36] and 1.8% of global greenhouse gas emissions.

Virtually all hydrogen produced for the current market is used in oil refining (40 MtH2 in 2021) and industry (54 MtH2) [6]. In oil refining, hydrogen is used, in a process known as hydrocracking, to convert heavy petroleum sources into lighter fractions suitable for use as fuels. Industrial uses mainly comprise ammonia production to make fertilizers (34 MtH2 in 2021), methanol production (15 MtH2) and the manufacture of direct reduced iron (5 MtH2)

5.3.1. Production

Hydrogen gas is produced by several industrial methods [37]. Nearly all of the world's current supply of hydrogen is created from fossil fuels [38, 39]. Most hydrogen is gray hydrogen made through steam methane reforming. In this pro cess, hydrogen is produced from a chemical reaction between steam and methane, the main component of natural gas. Producing one tonne of hydrogen through this process emits 6.6–9.3 tonnes of carbon dioxide [40]. When carbon capture and storage is used to remove a large fraction of these emissions, the product is known as blue hydrogen [41].

Green hydrogen is usually understood to be produced from renewable electricity via electrolysis of water [42, 43]. Less frequently, definitions of green hydrogen include hydrogen produced from other low-emission sources such as biomass [44]. Producing green hydrogen is currently more expensive than producing gray hydrogen, and the efficiency of energy conversion is inherently low [45]. Other methods of hydrogen production include biomass gasification, methane pyrolysis, and extraction of underground hydrogen [46, 47].

As of 2023, less than 1% of dedicated hydrogen production is low-carbon, i.e. blue hydrogen, green hydrogen, and hydrogen produced from biomass [48].

5.4. Green Methanol

Green methanol is a liquid fuel that is produced from combining carbon dioxide and hydrogen ($CO2 + 3 H2 \rightarrow CH3OH + H2O$) under pressure and heat with catalysts. It is a way to reuse carbon capture for recycling. Methanol can store hydrogen economically at standard outdoor temperatures and pressures, compared to liquid hydrogen and ammonia that need to use a lot of energy to stay cold in their liquid state [49]. In 2023 the Laura Maersk was the first container ship to run on methanol fuel [50]. Ethanol plants in the mid-west are a good place for pure carbon capture to combine with hydrogen to make green methanol, with abundant wind and nuclear energy in Iowa, Minnesota, and Illinois [51, 52]. Mixing methanol with ethanol could make methanol a safer fuel to use because methanol doesn't have a visible flame in the daylight and doesn't emit smoke, and ethanol has a visible light yellow flame [53-55]. Green hydrogen production of 70% efficiency and a 70% efficiency of methanol production from that would be a 49% energy conversion efficiency [56].

5.4.1. Uses

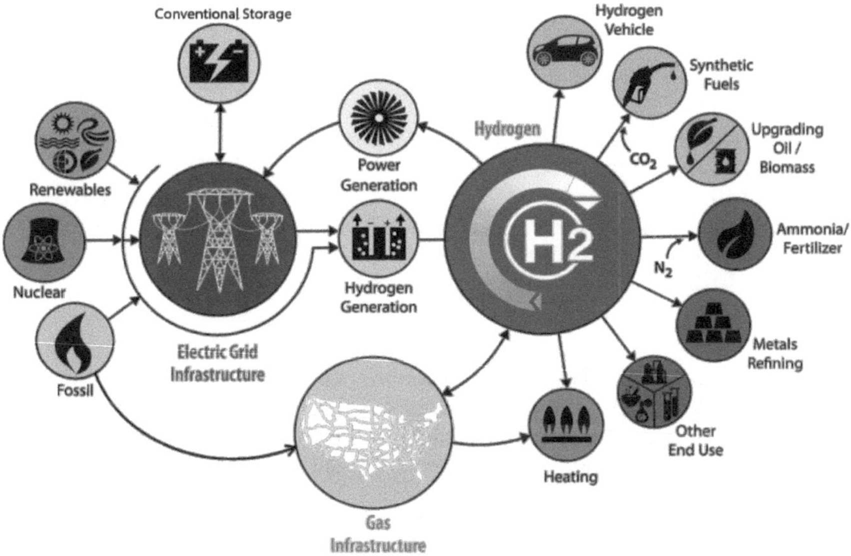

Some projected uses in the medium term, but analysts disagree [57]

Hydrogen can be deployed as a fuel in two distinct ways: in fuel cells which produce electricity, and via combustion to generate heat [58]. When hydrogen is consumed in fuel cells, the only emission at the point of use is water vapor [58]. Combustion of hydrogen can lead to the thermal formation of harmful nitrogen oxides emissions [58].

5.4.2. Industry

In the context of limiting global warming, low-carbon hydrogen (particularly green hydrogen) is likely to play an important role in decarbonizing industry [59]. Hydrogen fuel can produce the intense heat required for industrial production of steel, cement, glass, and chemicals, thus contributing to the de-carbonization of industry alongside other technologies, such as electric arc furnaces for steel-making. However, it is likely to play a larger role in providing industrial feedstock for cleaner production of ammonia and organic chemicals. For example, in steel-making, hydrogen could function as a clean energy carrier and also as a low-carbon catalyst replacing coal-derived coke [32].

The imperative to use low-carbon hydrogen to reduce greenhouse gas emissions has the potential to reshape the geography of industrial activities, as locations with appropriate hydrogen production potential in different regions will interact in new ways with logistics infrastructure, raw material availability, human and technological capital [59].

5.4.3. Transport

Much of the interest in the hydrogen economy concept is focused on hydrogen vehicles, particularly planes [60, 61]. Hydrogen vehicles produce significantly less local air pollution than conventional vehicles [62]. By **2050**, the energy requirement for transportation might be between 20% and 30% fulfilled by hydrogen and synthetic fuels [63-65].

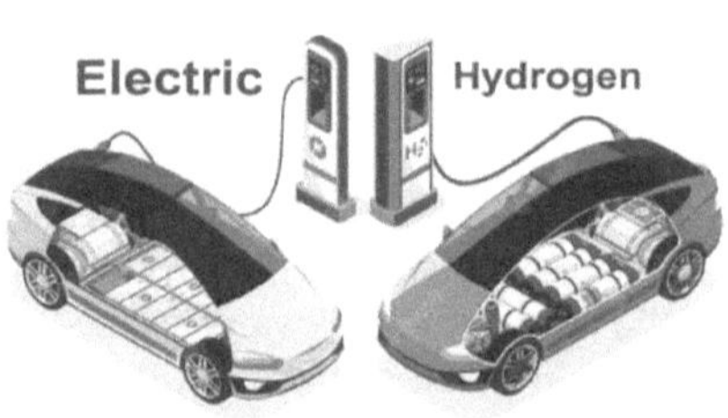

Hydrogen used to decarbonize transportation is likely to find its largest applications in shipping, aviation and to a lesser extent heavy goods vehicles, through the use of hydrogen-derived synthetic fuels such as ammonia and methanol, and fuel cell technology [8]. Hydrogen has been used in fuel cell buses for many years. It is also used as a fuel for spacecraft propulsion.

In the International Energy Agency's 2022 Net Zero Emissions Scenario (NZE), hydrogen is forecast to account for 2% of rail energy demand in 2050, while 90% of rail travel is expected to be electrified by then (up from 45% today). Hydrogen's role in rail would likely be focused on lines that prove difficult or costly to electrify [66]. The NZE foresees hydrogen meeting approximately 30% of heavy truck energy demand in 2050, mainly for long-distance heavy freight (with battery electric power accounting for around 60%) [67].

Although hydrogen can be used in adapted internal combustion engines, fuel cells, being electrochemical, have an efficiency advantage over heat engines. Fuel cells are more expensive to produce than common internal combustion engines but also require higher purity hydrogen fuel than internal combustion engines [68].

In the light road vehicle segment including passenger cars, by the end of 2022, 70,200 fuel cell electric vehicles had been sold worldwide, compared with 26 million plug-in electric vehicles [30]. With the rapid rise of electric vehicles and associated battery technology and infrastructure, hydrogen's role in cars is minuscule [27, 28].

5.4.4. Energy System Balancing and Storage

Green hydrogen, from electrolysis of water, has the potential to address the variability of renewable energy output. Hydrogen energy is a clean fuel that produces only water when used in fuel cells. While most hydrogen comes from natural gas, green hydrogen from renewables offers a zero-emission alternative. Producing green hydrogen can both reduce the need for renewable power curtailment during periods of high renewables output and be stored long-term to provide for power generation during periods of low output [69, 70].

5.5. Ammonia

An alternative to gaseous hydrogen as an energy carrier is to bond it with nitrogen from the air to produce ammonia, which can be easily liquefied, transported, and used (directly or indirectly) as a clean and renewable fuel [71, 72]. Among disadvantages of ammonia as an energy carrier are its high toxicity, energy efficiency of NH3 production from N2 and H2, and poisoning of PEM Fuel Cells by traces of non-decomposed NH3 after NH3 to N2 conversion.

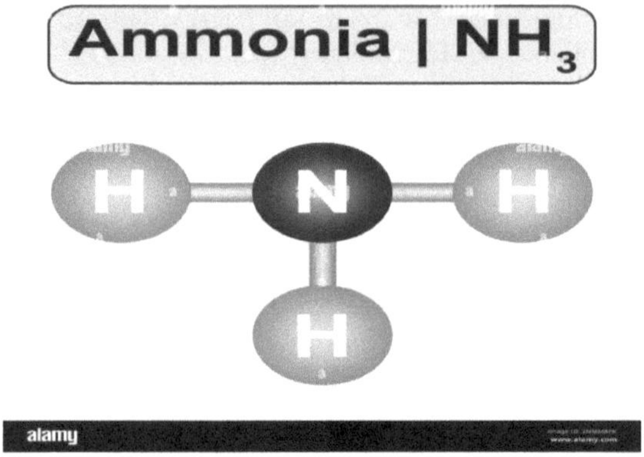

5.6. Buildings

Numerous industry groups (gas networks, gas boiler manufacturers) across the natural gas supply chain are promoting hydrogen combustion boilers for space and water heating, and hydrogen appliances for cooking, to reduce energy-related CO2 emissions from residential and commercial buildings [73, 74]. The proposition is that current end-users of piped natural gas can await the conversion of and supply of hydrogen to existing natural gas grids, and then swap heating and cooking appliances, and that there is no need for consumers to do anything now.

A review of 32 studies on the question of hydrogen for heating buildings, independent of commercial interests, found that the economics and climate benefits of hydrogen for heating and cooking generally compare very poorly with the deployment of district heating networks, electrification of heating (principally through heat pumps) and cooking, the use of solar thermal, waste heat and the installation of energy efficiency measures to reduce energy demand for heat [11]. Due to inefficiencies in hydrogen production, using blue hydrogen to replace natural gas for heating could require three times as much methane, while using green hydrogen would need two to three times as much electricity as heat pumps [11]. Hybrid heat pumps, which combine the use of an electric heat pump with a hydrogen boiler may play a role in residential heating in areas where upgrading networks to meet peak electrical demand would otherwise be costly [11].

The widespread use of hydrogen for heating buildings would entail higher energy system costs, higher heating costs and higher environmental impacts than the alternatives, although a niche role may be appropriate in specific contexts and geographies. If deployed, using hydrogen in buildings would drive up the cost of hydrogen for harder-to-decarbonize applications in industry and transport.

5.7. Bio-SNG

As of 2019 although technically possible production of syngas from hydrogen and carbon-dioxide from bio-energy with carbon capture and storage (BECCS) via the Sabatier reaction is limited by the amount of sustainable bioenergy available [75], therefore any bio-SNG made may be reserved for production of aviation biofuel [76].

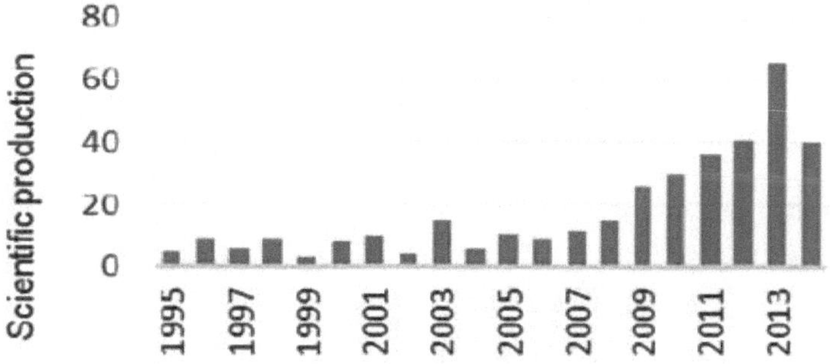

5.8. Safety

A NASA engineer sweeps an area with a corn broom to find the location of a hydrogen fire. Hydrogen burns with a nearly-invisible flame.

In hydrogen pipelines and steel storage vessels, hydrogen molecules are prone to react with metals, causing hydrogen embrittlement and leaks in the pipeline or storage vessel [77]. Since it is lighter than air, hydrogen does not easily accumulate to form a combustible gas mixture. However, even without ignition sources, high-pressure hydrogen leakage may cause spontaneous combustion and detonation.

Hydrogen is flammable when mixed even in small amounts with air. Ignition can occur at a volumetric ratio of hydrogen to air as low as 4%.[78] In approximately 70% of hydrogen ignition accidents, the ignition source cannot be found, and it is widely believed by scholars that spontaneous ignition of hydrogen occurs.

Hydrogen fire, while being extremely hot, is almost invisible, and thus can lead to accidental burns [79]. Hydrogen, like most gases, can cause asphyxiation in the absence of adequate ventilation [80].

5.9. Hydrogen Infrastructure

Hydrogen pipelines

A hydrogen infrastructure is the infrastructure of hydrogen pipeline transport, points of hydrogen production and hydrogen stations for distribution as well as the sale of hydrogen fuel [81], and thus a crucial prerequisite before a successful commercialization of fuel cell technology [82].

Hydrogen gasification plant for Belinka Perkemija [sl], 2015.

The hydrogen infrastructure would consist mainly of industrial hydrogen pipeline transport and hydrogen-equipped filling stations. Hydrogen stations which were not situated near a hydrogen pipeline would get supply via hydrogen tanks, compressed hydrogen tube trailers, liquid hydrogen trailers, liquid hydrogen tank trucks or dedicated onsite production.

Pipelines are the cheapest way to move hydrogen over long distances compared to other options. Hydrogen gas piping is routine in large oil-refineries, because hydrogen is used to hydrocrack fuels from crude oil. The IEA recommends existing industrial ports be used for production and existing natural gas pipelines for transport: also international co-operation and shipping [83].

South Korea and Japan [84], which as of 2019 lack international electrical interconnectors, are investing in the hydrogen economy [85]. In March 2020, the Fukushima Hydrogen Energy Research Field was opened in Japan, claiming to be the world's largest hydrogen production facility [86]. Much of the site is occupied by a solar array; power from the grid is also used for electrolysis of water to produce hydrogen fuel [87].

5.9.1. Storage

Several methods exist for storing hydrogen. These include mechanical approaches such as using high pressures and low temperatures, or employing chemical compounds that release H2 upon demand. While large amounts of hydrogen are produced by various industries, it is mostly consumed at the site of production, notably for the synthesis of ammonia. For many years hydrogen has been stored as compressed gas or cryogenic liquid, and transported as such in cylinders, tubes, and cryogenic tanks for use in industry or as propellant in space programs. The overarching challenge is the very low boiling point of H2: it boils around 20.268 K (−252.882 °C or −423.188 °F). Achieving such low temperatures requi0res expending significant energy.

Liquid hydrogen Tank at NASA Kennedy Space Center

Although molecular hydrogen has very high energy density on a mass basis, partly because of its low molecular weight, as a gas at ambient conditions it has very low energy density by volume. If it is to be used as fuel stored on board a vehicle, pure hydrogen gas must be stored in an energy-dense form to provide sufficient driving range. Because hydrogen is the smallest molecule, it easily escapes from containers. Its effective 100-year global warming potential (GWP100) is estimated to be 11.6 ± 2.8.

5.9.2. Power Plants

Xcel Energy is going to build two combined cycle power plants in the Midwest that can mix 30% hydrogen with the natural gas [88]. Intermountain Power Plant is being retrofitted to a natural gas/hydrogen power plant that can run on 30% hydrogen as well, and is scheduled to run on pure hydrogen by 2045 [89].

5.9.3. Costs

More widespread use of hydrogen in economies entails the need for investment and costs in its production, storage, distribution and use. Estimates of hydrogen's cost are therefore complex and need to make assumptions about the cost of energy inputs (typically gas and electricity), production plant and method (e.g. green or blue hydrogen), technologies used (e.g. alkaline or proton exchange membrane electrolysers), storage and distribution methods, and how different cost elements might change over time [90]. These factors are incorporated into calculations of the levelized costs of hydrogen (LCOH). The following table shows a range of estimates of the levelized costs of gray, blue, and green hydrogen, expressed in terms of US$ per kg of H2 (where data provided in other currencies or units, the average exchange rate to US dollars in the given year are used, and 1 kg of H2 is assumed to have a calorific value of 33.3kWh).

Production method	Note	Current cost (2020–2022)	Projected 2030 cost	Projected 2050 cost
Gray hydrogen (not including a carbon tax)				
International Energy Agency [91]	2022 costs estimated for June, when gas prices peaked in the wake of Russia's invasion of Ukraine	2021 1.0–2.5 2022 4.8–7.8	–	–
PWC [92]		2021 1.2–2.4		
Blue hydrogen				
International Energy Agency [91]	2022 costs estimated for June, when gas prices peaked in the wake of Russia's invasion of Ukraine	2021 1.5–3.0 2022 5.3–8.6	–	–
UK government [93]	Range dependent on gas price	2020 1.6–2.7	1.6–2.7	1.6–2.8
GEP [94]		2022 2.8–3.5	-	-

Energy Transitions Commission [90]		2020 1.5–2.4	1.3–2.3	1.4–2.2
		Green hydrogen		
International Energy Agency [91]	2030 and 2050 estimates are using solar power in regions with good solar conditions	2021 4.0–9.0 2022: 3.0-4.3	<1.5	<1.0
UK government [93]	Using grid electricity, UK specific; range dependent on electricity price, and electrolyser technology and cost	**2020** 4.9–7.9	4.4–6.6	4.0–6.3
	Using otherwise curtailed renewable electricity, UK specific; range dependent on electrolyser technology and cost	2020 2.4–7.9	1.7–5.6	1.5–4.6
IRENA [95]		2020 2.2–5.2	1.4–4.1	1.1–3.4
GEP [94]	Source notes green H2 production cost has fallen by 60% since 2010	2022 3.0–6.0		
Lazard [96]		2022 2.8–5.3		
PWC [92]		2021 3.5–9.5	1.8–4.8	1.2–2.4
Energy Transitions Commission [90]		2020 2.6–3.6	1.0–1.7	0.7–1.2

The range of cost estimates for commercially available hydrogen production methods is broad. As of 2022, gray hydrogen is cheapest to produce without a tax on its CO_2 emissions, followed by blue and green hydrogen. Blue hydrogen production costs are not anticipated to fall substantially by 2050 [93], can be expected to fluctuate with natural gas prices and could face carbon taxes for uncaptured emissions [90]. The cost of electrolysers fell by 60% from 2010 to 2022 [94], before rising slightly due to an increasing cost of capital. Their cost is projected to fall significantly to 2030 and 2050 [97], driving down the cost of green hydrogen alongside the falling cost of renewable power generation [98, 99]. It is cheapest to produce green hydrogen with surplus renewable power that would otherwise be curtailed, which favors electrolyzers capable of responding to low and variable power levels [97].

A 2022 Goldman Sachs analysis anticipates that globally green hydrogen will achieve cost parity with grey hydrogen by 2030, earlier if a global carbon tax is placed on gray hydrogen. In terms of cost per unit of energy, blue and gray hydrogen will always cost more than the fossil fuels used in its production, while green hydrogen will always cost more than the renewable electricity used to make it.

Subsidies for clean hydrogen production are much higher in the US and EU than in India [99].

5.9.3. Examples and Pilot Programs

A Mercedes-Benz O530 Citaro powered by hydrogen fuel cells in Brno, Czech Republic

The distribution of hydrogen for the purpose of transportation is being tested around the world, particularly in the US (California, Massachusetts), Canada, Japan, the EU (Portugal, Norway, the Netherlands, Denmark, Germany), and Iceland.

An indicator of the presence of large natural gas infrastructures already in place in countries and in use by citizens is the number of natural gas vehicles present in the country. The countries with the largest amount of natural gas vehicles are (in order of magnitude) [100] Iran, China, Pakistan, Argentina, India, Brazil, Italy, Colombia, Thailand, Uzbekistan, Bolivia, Armenia, Bangladesh, Egypt, Peru, Ukra ine, the United States. Natural gas vehicles can also be converted to run on hydrogen.

Also, in a few private homes, fuel cell micro-CHP plants can be found, which can operate on hydrogen, or other fuels as natural gas or LPG [101, 102].

5.9.3.1. Australia

Western Australia's Department of Planning and Infrastructure operated three Daimler Chrysler Citaro fuel cell buses as part of its Sustainable Transport Energy for Perth Fuel Cells Bus Trial in Perth from 2004 to 2007 [103].

By November 2020 the Australian Renewable Energy Agency (ARENA) had invested $55 million in 28 hydrogen projects, from early stage research and development to early stage trials and deployments. The agency's stated goal is to produce hydrogen by electrolysis for $2 per kilogram, announced by Minister for Energy and Emissions Angus Taylor in a 2021 Low Emissions Technology Statement [104].

In October 2021, Queensland Premier Annastacia Palaszczuk and private investor Andrew Forrest announced that Queensland would be home to the world's largest hydrogen plant [105].

In November 2024, the South Australian Government's plan to spend A$593 million on a 200MW hydrogen energy plant near Whyalla was granted federal approval under the EPBC Act. The project is planned to be developed by the Office of Hydrogen Power SA, ATCO Australia, and BOC [106], and intended "to supply additional grid stability for homes and businesses around the state, by using excess renewable energy generated from large-scale wind and solar farms to provide a consistent output of supply". Construction was scheduled to start in 2025, with completion and commissioning happening in 2026 [107].

5.9.3.2. European Union

Countries in the EU which have a relatively large natural gas pipeline system already in place include Belgium, Germany, France, and the Netherlands [108]. In 2020, The EU launched its European Clean Hydrogen Alliance (ECHA) [109, 110].

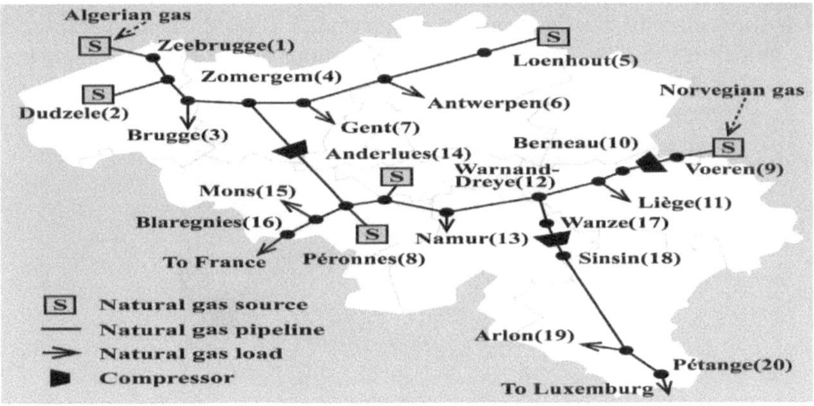

5.9.3.3. France

Green hydrogen has become more common in France. A €150 million Green Hydrogen Plan was established in 2019, and it calls for building the infrastructure necessary to create, store, and distribute hydrogen as well as using the fuel to power local transportation systems like buses and trains. Corridor H2, a similar initiative, will create a network of hydrogen distribution facilities in Occitania along the route between the Mediterranean and the North Sea. The Corridor H2 project will get a €40 million loan from the EIB [111, 112].

5.9.3.4. Germany

German car manufacturer BMW has been working with hydrogen for years. [113]. The German government has announced plans to hold tenders for 5.5 GW of new hydrogen-ready gas-fired power plants and 2 GW of "comprehensive H2-ready modernizations" of existing gas power stations at the end of 2024 or beginning of 2025 [114].

5.9.3.5. Iceland

Iceland has committed to becoming the world's first hydrogen economy by the year 2050 [115]. Iceland is in a unique position. Presently, it imports all the petroleum products necessary to power its automobiles and fishing fleet. Iceland has large geothermal resources, so much that the local price of electricity actually is lower than the price of the hydrocarbons that could be used to produce that electricity.

Iceland already converts its surplus electricity into exportable goods and hydrocarbon replacements. In 2002, it produced 2,000 tons of hydrogen gas by electrolysis, primarily for the production of ammonia (NH3) for fertilizer. Ammonia is produced, transported, and used throughout the world, and 90% of the cost of ammonia is the cost of the energy to produce it.

Neither industry directly replaces hydrocarbons. Reykjavík, Iceland, had a small pilot fleet of city buses running on compressed hydrogen [116], and research on powering the nation's fishing fleet with hydrogen is under way (for example by companies as Icelandic New Energy). For more practical purposes, Iceland might process imported oil with hydrogen to extend it, rather than to replace it altogether. The Reykjavík buses are part of a larger program, HyFLEET:CUTE [117], operating hydrogen fueled buses in eight European cities. HyFLEET:CUTE buses were also operated in Beijing, China and Perth, Australia. A pilot project demon-strating a hydrogen economy is operational on the Norwegian island of Utsira. The installation combines wind power and hydrogen power. In periods when there is surplus wind energy, the excess power is used for generating hydrogen by elect- rolysis. The hydrogen is stored, and is available for power generation in periods when there is little wind.

5.9.3.6. India

India is said to adopt hydrogen and H-CNG, due to several reasons, amongst which the fact that a national rollout of natural gas networks is already taking place and natural gas is already a major vehicle fuel. In addition, India suffers from extreme air pollution in urban areas [118, 119]. According to some estimates, nearly 80% of India's hydrogen is projected to be green, driven by cost declines and new production technologies [120].

Currently however, hydrogen energy is just at the Research, Development and Demonstration (RD&D) stage [121, 122]. As a result, the number of hydrogen stations may still be low [123], although much more are expected to be introduced soon [124-126].

5.9.3.7. Poland

It planning open first hydrogen publication stations, The Ministry of Climate and Environment (MKiŚ) will soon schan competitions for 2-3 hydrogen refueling stations, Polish Deputy Minister in this ministry Krzysztof Bolesta [127].

5.9.3.8. Saudi Arabia

Saudi Arabia as a part of the NEOM project, is looking to produce roughly 1.2 million tonnes of green ammonia a year, beginning production in 2025 [128].

In Cairo, Egypt, Saudi real estate funding skyscraper project powered by hydrogen [129].

5.9.3.9. South Korea

The Ulsan Green Hydrogen Town is a hydrogen city project being developed in Ulsan. As of October 2024, 188km of underground pipelines have been laid to connect hydrogen produced as a byproduct from petrochemical complexes to the city center [130].

5.9.3.10. Turkey

The Turkish Ministry of Energy and Natural Resources and the United Nations Industrial Development Organization created the International Centre for Hydrogen Energy Technologies (UNIDO-ICHET) in Istanbul in 2004 and it ran to 2012 [131]. In 2023 the ministry published Hydrogen Technologies Strategy and Roadmap [132].

5.9.3.11. United Kingdom

The UK started a fuel cell pilot program in January 2004, the program ran two Fuel cell buses on route 25 in London until December 2005, and switched to route RV1 until January 2007 [133]. The Hydrogen Expedition is currently working to create a hydrogen fuel cell-powered ship and using it to circumnavigate the globe, as a way to demonstrate the capability of hydrogen fuel cells [134]. In August 2021 the UK Government claimed it was the first to have a Hydrogen Strategy and produced a document [135].

In August 2021, Chris Jackson quit as chair of the UK Hydrogen and Fuel Cell Association, a leading hydrogen industry association, claiming that UK and Norwegian oil companies had intentionally inflated their cost projections for blue hydrogen in order to maximize future technology support payments by the UK government.[136]

5.9.3.12. United States

Several domestic U.S. automobile companies have developed vehicles using hydrogen, such as GM and Toyota [137]. However, as of February 2020, infrastructure for hydrogen was underdeveloped except in some parts of California [138]. The United States have their own hydrogen policy. A joint venture between NREL and Xcel Energy is combining wind power and hydrogen power in the same way in Colorado [139]. Hydro in Newfoundland and Labrador are converting the current wind-diesel Power System on the remote island of Ramea into a Wind-Hydrogen Hybrid Power Systems facility [140]. Five pump station hubs being delivered to heavy-duty H2 trucks in Texas [141]. Hydrogen City built Green by Hydrogen International (GHI), to planning open in 2026 [142].

In 2006, Florida's infrastructure project was commissioned [143]. First opened Orlando as public bus transportation, Ford Motor Company announced putting a fleet of hydrogen-fueled Ford E-450 [144, 145]. Liquidated hydrogen mobile system was constructed at Titusville [146, 147]. An FPL's pilot clean hydrogen facility operated in Okeechobee County [148].

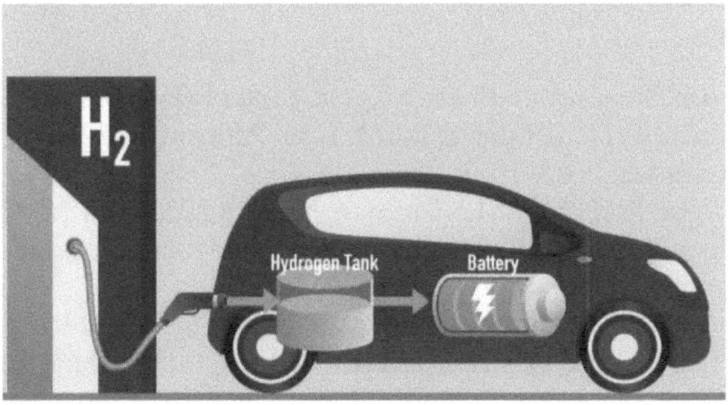

A similar pilot project on Stuart Island uses solar power, instead of the wind power, to generate electricity. When excess electricity is available after the batteries are fully charged, hydrogen is generated by electrolysis and stored for later production of electricity by fuel cell [149]. The US also has a large natural gas pipeline system already in place [108],

5.9.3.13. Vietnam

Việt Nam Energy Association has included green hydrogenation support [150]. Australian clean energy company Pure Hydrogen Corporation Limited announced on July 2025 that it has signed an MoU with Vietnam public transportation [151].

5.10. References

[1]. International Renewable Energy Agency (2022-03-29). "World Energy Transitions Outlook 1-5C Pathway 2022 edition". IRENA. p. 227. Retrieved 2023-10-06.

[2]. Yap, Jiazhen; McLellan, Benjamin (6 January 2023). "A Historical Analysis of Hydrogen Economy Research, Development, and Expectations, 1972 to 2020". Environments. 10 (1): 11.

[3]. Greenhouse gas emissions totalled 49.3 Gigatonnes CO2e in 2021. "Global Greenhouse Gas Emissions: 1990–2020 Preliminary 2021". Rhodium Group. 19 December 2022. Retrieved 2023-09-21.

[4]. "Hydrogen". IEA. 10 July 2023. "Energy" section. Retrieved 2023-09-21.

[5]. "Hydrogen". IEA. Retrieved 2024-03-24.

[6]. IEA (2022). Global Hydrogen Review 2022. International Energy Agency. Retrieved 2023-08-25.

[7]. "Hydrogen could be used for nearly everything. It probably shouldn't be". MIT Technology Review. Retrieved 2024-05-13.

[8]. IPCC (2022). Shukla, P.; Skea, J.; Slade, R.; Al Khourdajie, A.; et al. (eds.). Climate Change 2022: Mitigation of Climate Change (PDF). Contribution of Working Group III to the Sixth Assessment Report of the Intergovernmental Panel on Climate Change. Cambridge, UK and New York, NY, US: Cambridge University Press (In Press). pp. 91–92.

[9]. IRENA (2021). "World Energy Transitions Outlook: 1.5 °C Pathway". International Renewable Energy Agency. Abu Dhabi. p. 95. Retrieved 2023-09-21.

[10]. Plötz, Patrick (2022-01-31). "Hydrogen technology is unlikely to play a major role in sustainable road transport". Nature Electronics. 5 (1): 8–10. ISSN 2520-1131. S2CID 246465284.

[11]. Rosenow, Jan (September 2022). "Is heating homes with hydrogen all but a pipe dream? An evidence review". Joule. 6 (10): 2225–2228.

[12]. Barnard, Michael (2023-10-22). "What's New On The Rungs Of Liebreich's Hydrogen Ladder?". CleanTechnica. Retrieved 2024-02-17.

[13]. Goldman Sachs Research. "Carbonomics: Clean Hydrogen Revolution". Goldman Sachs. pp. 4–6. Retrieved 2023-09-25.

[14]. "Hydrogen Hotshot Initiative". DOE. 31 August 2021.

[15]. "Daedalus or Science and the Future, A paper read to the Heretics, Cambridge, on February 4th, 1923 – Transcript 1993". Archived on 2017-11-15. Retrieved 2016-01-16.

[16]. National Hydrogen Association; United States Department of Energy.

"The History of Hydrogen". hydrogenassociation.org. National Hydrogen Association. p. 1. Archived on 14 July 2010. Retrieved 17 December 2010.
[17]. Bockris, J. O'M. (1972-06-23). "A Hydrogen Economy". Science. 176 (4041): 1323. ISSN 0036-8075. PMID 17820918.
[18]. Jones, Lawrence W (13 March 1970). Toward a liquid hydrogen fuel economy. University of Michigan Environmental Action for Survival Teach In. Ann Arbor, Michigan: University of Michigan. hdl:2027.42/5800.
[19]. Jones, Lawrence W. (March 13, 1970). Toward a Liquid Hydrogen Fuel Economy (PDF). pp. 2–3.
[20]. IRENA (2022), Geopolitics of the Energy Transformation: The Hydrogen Factor, International Renewable Energy Agency, Abu Dhabi. ISBN 978-92- 9260-370-0.
[21]. Bakker, S. (2010). "The car industry and the blow-out of hydrogen hype". Energy Policy. 38 (11): 6540–6544. Archived on 2018-11-03. Retrieved 2019-12-11.
[22]. Harrison, J. "Reactions: Hydrogen hype". Chemical Engineer. 58: 774–775. Archived on 2021-02-08. Retrieved 2017-08-31.
[23]. Rizzi, Francesco A., Eleonora Liberati, Guglielmo Frey, Marco (2014). "Technological trajectories in the automotive industry: are hydrogen technologies still a possibility?". Jour. of Cleaner Production. 66: 328–336.
[24]. "Can a viable industry emerge from the hydrogen shakeout?". The Economist. ISSN 0013-0613. Retrieved 2023-09-26.
[25]. Murai, Shusuke (2018-03-05). "Japan's top auto and energy firms tie up to promote development of hydrogen stations". The Japan Times Online. Japan Times. Archived on 2018-04-17. Retrieved 16 April 2018.
[26]. Mishra, Ankit (2018-03-29). "Prospects of fuel-cell electric vehicles boosted with Chinese backing". Energy Post. Archived on 2018-04-17. Retrieved 16 April 2018.
[27]. Plötz, Patrick (January 2022). "Hydrogen technology is unlikely to play a major role in sustainable road transport". Nature Electronics. 5 (1): 8–10. ISSN 2520-1131.
[28]. Collins (l_collins), Leigh (2022-02-02). "'Hydrogen unlikely to play major role in road transport, even for heavy trucks': Fraunhofer". Recharge | Latest Renewable Energy News. Retrieved 2023-09-08.
[29]. Chu, Yidan; Cui, Hongyang. Annual update on the global transition to electric vehicles: 2022. International Council on Clean Transportation. pp. 2–3. Retrieved 2023-08-25.

[30]. Global EV Outlook 2023. IEA. 26 April 2023. pp. 14–24. Retrieved 2023-08-25.
[31]. Kjellberg-Motton, Brendan (2022-02-07). "Steel decarbonisation gathers speed | Argus Media". www.argusmedia.com. Retrieved 2023-09-07.
[32]. Blank, Thomas; Molly, Patrick (January 2020). "Hydrogen's Decarbonization Impact for Industry". Rocky Mountain Institute. pp. 2, 7, 8. Archived on 22 September 2020.
[33]. "H2-Generation Market Size, Share & Trends Analysis Report, – 2030". www.grandviewresearch.com. Retrieved 2023-08-30.
[34]. "Executive summary – Global Hydrogen Review 2022 – Analysis". IEA. Retrieved 2023-09-21.
[35]. "Hydrogen". IEA. Retrieved 2023-09-21.
[36]. Energy-related emissions totalled 36.3 Gigatonnes CO2 in 2021. "Global CO2 emissions rebounded to their highest history in 2021 – News". IEA. 8 March 2022. Retrieved 2023-09-21.
[37]. Fan, Lixin; Tu, Zhengkai; Chan, Siew Hwa (2021). "Recent development of hydrogen and fuel cell technologies: A review". Energy Reports. 7: 8421–8446.
[38]. Reed, Stanley; Ewing, Jack (13 July 2021). "Hydrogen Is One Answer to Climate Change. Getting It Is the Hard Part". The New York Times.
[39]. Rosenow, Jan (27 September 2022). "Is heating homes with hydrogen all but a pipe dream? An evidence review". Joule. 6 (10): 2225–2228.
[40]. Bonheure, Mike; et al. (March 2021). "Dream or Reality? Electrification of the Chemical Process Industries". CEP Magazine. American Institute of Chemical Engineers. Archived on 17 July 2021. Retrieved 6 July 2021.
[41]. Griffiths, Steve; et al. (October 2021). Energy Research & Social Science. 80:102208. Bibcode:2021ERSS...8002208G.
[42]. Squadrito, Gaetano; Maggio, Gaetano; Nicita, Agatino (November 2023). "The green hydrogen revolution". Renewable Energy. 216: 119041.
[43]. Deign, Jason (2020-06-29). "So, What Exactly Is Green Hydrogen?". Greentechmedia. Archived on 2022-03-23. Retrieved 2022-02-11.
[44]. Squadrito, G.; Maggio, Gaetano; Nicita, Agatino (November 2023). "The green hydrogen revolution". Renewable Energy. 216: 119041.
[45]. Evans, Simon; Gabbatiss, Josh (30 November 2020). "In-depth Q&A: Does the world need hydrogen to solve climate change?". Carbon Brief. Archived on 1 December 2020. Retrieved 1 December 2020.
[46]. "Natural Hydrogen: A Potential Clean Energy Source Beneath Our Feet".

Yale E360. Retrieved 2024-03-23.
[47]. Hassanpouryouzband, A.; Wilkinson, M.; Haszeldine, R Stuart (2024). "Hydrogen energy futures – foraging or farming?". Chemical Society Reviews. 53 (5): 2258–2263.
[48]. "Hydrogen". IEA. 10 July 2023. "Energy" section. Retrieved 2023-09-21.
[49]. Song, Qianqian; et al. (2022-09-01). "A comparative study on energy efficiency of the maritime supply chains for liquefied hydrogen, ammonia, methanol and natural gas". Carbon Capture Science & Technology. 4: 100056.
[50]. "World's 'first green container ship' christened in Denmark". euronews. 2023-09-14. Retrieved 2024-08-14.
[51]. Strong, Jared (2024-02-17). "Green methanol: Carbon dioxide pipeline alternative• Nebraska Examiner". Nebraska Examiner. Retrieved 2024-08-14.
[52]. Cordero-Lanzac, et al. (2022-05-01). "A techno-economic and life cycle assessment for the production of green methanol from CO2: catalyst and process bottlenecks". Journal of Energy Chemistry. 68: 255–266.
[53]. Li, Shu-hao; et al. (2022). "Effects of Ethanol and Methanol on the Combustion Characteristics of Gasoline with the Revised Variation Disturbance Method". ACS Omega. 7 (21): 17797–17810.
[54]. "The Horror of Methanol Fires | Last Moments". YouTube. 17 March 2023.
[55]. "Isopropanol blended with aqueous ethanol for flame coloration without use of salts or hazardous solvents".
[56]. "Green Methanol Production-A Techno-Economic Analysis". www.linkedin.com. Retrieved 2024-08-14.
[57]. Barnard, Michael (2023-10-22). "What's New On The Rungs Of Liebreich's Hydrogen Ladder?". CleanTechnica. Retrieved 2024-03-10.
[58]. Lewis, Alastair C. (10 June 2021). "Optimising air quality co-benefits in a hydrogen economy: for hydrogen-specific standards for NO x emissions". Environmental Science: Atmospheres. 1 (5): 201–207.
[59]. IPCC (2022). Shukla, P.R.; et al. Climate Change 2022: Mitigation of Climate Change. Contribution of Working Group III 6th Assessment Report of the Intergovernmental Panel on Climate Change. Cambridge, UK and New York, NY, US: Cambridge University Press p. 1184.
[60]. "Is the time now ripe for planes to run on hydrogen?". The Economist. ISSN 0013-0613. Retrieved 2024-02-17.
[61]. Yusaf, Talal; et al. (2024-01-02).

"Sustainable hydrogen energy in aviation – A narrative review". International Journal Hydrogen Energy. 52: 1026–1045. ISSN 0360-3199.
[62]. "This company have solved one of the hardest problems in clean energy". Vox. 2018-02-16. Archived on 2019-11-12. Retrieved 9 February 2019.
[63]. IRENA. "The Hydrogen Factor". irena.org. Archived on 2022-10-19. Retrieved 2022-10-19.
[64]. "Sustainable fuels and their role in decarbonizing energy | McKinsey". www.mckinsey.com. Retrieved 2022-10-19.
[65]. Spiryagin, Maksym; Dixon, Roger; Oldknow, Kevin; Cole, Colin (2021). "Preface to special issue on hybrid H2-technologies for railway operations". Railway Engineering Science. 29 (3): 211. ISSN 2662-4753.
[66]. World energy outlook 2022. International Energy Agency. 27 October 2022. p. 150.
[67]. Cozzi, Laura; Gould, Tim. World Energy Outlook 2022. International Energy Agency. p. 148.
[68]. Stępień, Zbigniew (January 2021). "A Comprehensive Overview of Hydrogen-Fueled Internal Combustion Engines: Achievements and Future Challenges". Energies. 14 (20): 6504. ISSN 1996-1073.
[69]. Schrotenboer, Albert H.; et al. (2022). "A Green Hydrogen Energy System: Optimal control strategies for integrated hydrogen storage and power generation with wind energy". Renewable and Sustainable Energy Review. 168: 112744.
[70]. Lipták, Béla (2022). "Hydrogen is key to sustainable green energy". Control. Retrieved February 12, 2023.
[71]. Agosta, Vito (July 10, 2003). "The Ammonia Economy". Archived on May 13, 2008. Retrieved 2008-05-09.
[72]. "Renewable Energy". Iowa Energy Center. Archived on 2008-05-13. Retrieved 2008-05-09.
[73]. Collins, Leigh (2021-12-10). "Even the European gas lobby can't make a case for hydrogen boilers — so why does it say gases are needed to decarbonise heating?". Recharge | Latest renewable energy news. Retrieved 2023-09-25.
[74]. Roth, Sammy (2023-02-09). "California declared war on natural gas. Now the fight is going national". Los Angeles Times. Retrieved 2023-09-25.
[75]. UKCCC H2 2018, p. 79: The potential for bio-gasification with CCS to be deployed scale limited by the amount of sustainable bioenergy available. ."
[76]. UKCCC H2 2018, p. 33: production of biofuels, even with CCS, is only one of the best uses of the finite sustainable bio-resource if the fossil fuels it

displaces cannot otherwise feasibly be displaced (e.g. use of biomass to produce aviation biofuels with CCS)."
[77]. Li, Hao; et al. (2022-11-01). "Safety of hydrogen storage and transportation: An overview on mechanisms, techniques, and challenges". Energy Reports. 8: 6258–6269. doi:10.1016/j.egyr.2022.04.067. ISSN 2352-4847.
[78]. Yang, Fuyuan; et al. (2021-09-03). "Review on hydrogen safety issues: Incident statistics, hydrogen diffusion, and detonation process". International Journal of Hydrogen Energy. 46 (61): 31467–31488. ISSN 0360-3199.
[79]. "Hydrogen Detection Tape Saves Time and Lives | NASA Spinoff". spinoff.nasa.gov. 2016. Retrieved 2025-02-23.
[80]. U.S. Department of Energy. "Current Safe Operating Practices". Energy.gov. Retrieved 2025-02-24.
[81]. "Hydrogen infrastructure project launches in USA". 14 May 2013.
[82]. Eberle, Ulrich; et al. "Fuel cell electric vehicles and hydrogen infrastructure: status 2012".
Energy & Environmental Science. Retrieved 23 December 2014.
[83]. IEA H2 2019, p. 15
[84]. "Japan's H2-Strategy and Its Economic and Geopolitical Implications". Etudes de l'Ifri. Archived on 10 February 2019. Retrieved 9 February 2019.
[85]. "South Korea's Hydrogen Economy Ambitions".
The Diplomat. Archived on 9 February 2019. Retrieved 9 February 2019.
[86]. "The world's largest-class hydrogen production, Fukushima Hydrogen Energy Research Field now is completed at Namie town in Fukushima". Toshiba Energy Press Releases. Toshiba Energy Systems and Solutions Corporations. 7 March 2020. Archived on 22 April 2020. Retrieved 1 April 2020.
[87]. Patel, Sonal (2022-07-01). "Fukushima Hydrogen Energy Research Field Demonstrates Hydrogen Integration". POWER Magazine.
Retrieved 2023-10-05.
[88]. Orenstein, Walker (2024-02-01).
"Xcel Energy wants Prairie Island nuclear facility, add two gas plants". www.startribune.com. Retrieved 2024-08-14.
[89]. "Chevron joins Mitsubishi in 300 GWh hydrogen storage project as construction continues". Utility Dive. Retrieved 2024-08-14.
[90]. Making the Hydrogen Economy Possible: Accelerating Clean Hydrogen in An Electrified Economy. Energy Transitions Commission. April 2021. Retrieved 2023-08-25.
[91]. Global Hydrogen Review 2022. IEA. 22 September 2022. p. 93.

Retrieved 2023-08-25.
[92]. PricewaterhouseCoopers. "Green hydrogen economy – predicted Development of tomorrow". PwC. Retrieved 2023-08-25.
[93]. "Hydrogen Production Costs 2021 annex: Key assumptions and outputs for production technologies". GOV.UK. Retrieved 2023-08-25.
[94]. Saini, Anshuman (January 12, 2023). "Green & Blue Hydrogen: Current Levelized Cost of Production & Outlook | GEP Blogs". www.gep.com. Retrieved 2023-08-25.
[95]. IRENA (2020), Green Hydrogen Cost Reduction: Scaling up Electrolysers to Meet the 1.5 °C Climate Goal, International Renewable Energy Agency, Abu Dhabi, p. 91.
[96]. 2023 Levelized Cost Of Energy+. Lazard. April 12, 2023. p. 27. Retrieved 2023-08-25.
[97]. Patonia, Aliaksei; Poudineh, Rahmat (January 2022). Cost-competitive green hydrogen: how to lower the cost of electrolysers?. Oxford Institute for Energy Studies. Retrieved 2023-08-25.
[98]. Roser, Max (2023-09-01). "Why did renewables become so cheap so fast?". Our World in Data.
[99]. Martin, Polly (2023-06-29). "India to offer green hydrogen production subsidy of up to $0.60/kg — for three years only". Hydrogen news and intelligence | Hydrogen Insight. Retrieved 2023-09-26.
[100]. "Worldwide NGV statistics". Archived on 2015-02-06. Retrieved 2019-09-29.
[101]. "Fuel Cell micro CHP". Archived on 2019-11-06. Retrieved 2019-10-23.
[102]. "Fuel cell micro Cogeneration". Archived on 2019-10-23. Retrieved 2019-10-23.
[103]. "Perth Fuel Cell Bus Trial". Department for Planning and Infrastructure, Government of Western Australia. 13 April 2007. Archived on 7 June 2008. Retrieved 2008-05-09.
[104]. "Australia's pathway to $2 per kg hydrogen". Australian Renewable Energy Agency. 30 November 2020. Archived on 2020-12-15. Retrieved 2021-01-06.
[105]. "'Green industrial revolution': Queensland announces plans to mass produce green ammonia". ABC News. October 11, 2021. Archived on 2021-10-12. Retrieved 2021-10-12 – via abc.net.au.
[106]. Howland, Nadia (3 November 2024). "South Australia's 200MW Whyalla hydrogen plant approved". Energy Source & Distribution. Retrieved 22 February 2025.
[107]. Howland, Nadia (19 February 2025). "Steelworks administration fuels doubt

over Whyalla hydrogen". Energy Source & Distribution. Retrieved 22 February 2025.
[108]. "Hydrogen transport & distribution". Archived on 2019-09-29. Retrieved 2019-09-29.
[109]. Pollet, Mathieu (2020). "AExplainer: EU Commission betting on hydrogen for a greener future?". euronews. Archived from the original on 2020-08-07. Retrieved 2020-08-14.
[110]. "ECHA". Archived on 2020-08-12. Retrieved 2020-08-14.
[111]. "French port bets big on floating wind farms planned in Mediterranean". European Investment Bank. Retrieved 2022-09-26.
[112]. "Green Hydrogen: A key investment for the energy transition". blogs.worldbank.org. 23 June 2022. Retrieved 2022-09-26.
[113]. "E3B1C256-BFCB-4CEF-88A6-1DCCD7666635". 24 October 2007. Archived on 2021-10-29. Retrieved 2021-10-12.
[114]. "Germany to tender for 5.5GW of new hydrogen-ready gas-fired power plants and 2GW of conversions". 8 July 2024.
[115]. Hannesson, Hjálmar W. (2007). "Climate change as a global challenge". Iceland Ministry for Foreign Affairs. Archived on 2014-01-07. Retrieved 2008-05-09.
[116]. Doyle, A. (2005). "Iceland's hydrogen buses zip toward oil-free economy". Reuters. Archived on July 24, 2012. Retrieved 2008-05-09.
[117]. "HyFLEET:CUTE?". Archived on 2008-02-24. Retrieved 2008-05-09.
[118]. "Hydrogen vehicles and refueling infrastructure in India". Archived on 2017-06-12. Retrieved 2019-09-28.
[119]. Das, L (1991). "Exhaust emission characterization of hydrogen-operated engine system: Nature of pollutants and their control techniques". International Journal of Hydrogen Energy. 16 (11): 765–775.
[120]. "UK-India Energy Collaborations report" (PDF).
[121]. "MNRE: FAQ". Archived on 2019-09-21. Retrieved 2019-09-28.
[122]. Overview of Indian Hydrogen Programme
[123]. "H2 stations worldwide". Archived on 2019-09-21. Retrieved 2019-09-28.
[124]. "India working on more H2 stations". 23 February 2016. Archived on 2019-09-21. Retrieved 2019-09-28.
[125]. "Shell plans to open 1200 fuel stations in India, some of which may include H2 refilling". The Economic Times. Archived on 2019-09-22. Retrieved 2019-09-28.
[126]. "Hydrogen Vehicles and Refueling Infrastructure in India". Archived on 2017-06-12. Retrieved 2019-09-28.

[127]. "ORLEN will Build the First Hydrogen Refueling Stations in Poland". 6 May 2021.
[128]. "Saudi Arabia's $5bn green hydrogen-based ammonia plant to begin production in 2025". Energy & Utilities. 21 April 2021. Archived on 2021-04-21. Retrieved 2022-01-13.
[129]. https://www.reuters.com/sustainability/saudi-firm-plans-hydrogen-powered-skyscraper-egypts-new-capital-2024-08-14/ [bare URL]
[130]. "'탄소제로' 주거·교통...울산형 수소도시 뜬다". Korea Economic Daily. 28 ctober 2024.
[131]. "Independent Mid-Term Review of the UNIDO Project: Establishment and operation of the International Centre for Hydrogen Energy Technologies (ICHET), TF/INT/03/002" (PDF). UNIDO. 31 August 2009. Archived on 1 June 2010. Retrieved 20 July 2010.
[132]. "Announcement–Türkiye Ministry of Energy and Natural Resources". enerji.gov.tr. Retrieved 2024-02-14.
[133]. "Hydrogen buses". Transport for London. Archived on March 23, 2008. Retrieved 2008-05-09.
[134]. "The Hydrogen Expedition" (PDF). January 2005. Archived on 2008-05-27. Retrieved 2008-05-09.
[135]. "UK Hydrogen Strategy" (PDF). UK Government. August 2021. Archived on 2021-08-19. Retrieved 2021-08-19.
[136]. Ambrose, Jillian (20 August 2021). "Oil firms made 'false claims' on blue hydrogen costs, says ex-lobby boss". The Guardian. London, United Kingdom. ISSN 0261-3077. Archivedon 2021-08-24. Retrieved 2021-08-24.
[137]. "Are hydrogen fuel cell vehicles the future of autos?". ABC News. Archived on 2021-01-17. Retrieved 2021-01-18.
[138]. Siddiqui, Faiz. "The plug-in electric car is having its moment. But despite false starts, Toyota is still trying to make the fuel cell happen". Washington Post. ISSN 0190-8286. Archived on 2021-01-19. Retrieved 2021-01-18.
[139]. "Experimental 'wind to hydrogen' system up and running". Physorg.com. January 8, 2007. Archived on 2013-01-26. Retrieved 2008-05-09.
[140]. "Hydrogen Engine Center Receives Order for Hydrogen Power Generator 250kW Generator for Wind/Hydrogen Demonstration". Hydrogen Engine Center, May 16, 2006. Archived on May 27, 2008. Retrieved 2008-05-09.
[141]. Kilgore, Erin (2024-01-12). "Texas Hydrogen Stations Infrastructure Gets Boost From Biden Administration". Hydrogen Fuel News.
[142]. Blain, Loz (2022-03-08). "World's largest green H2 hub, Hydrogen City, to open in Texas in 2026". New Atlas.
[143]. "The Florida Hydrogen Initiative". Hydrogen Program. April 12, 2007.

[144]. "First Hydrogen Station Opens". Tampa Bay Times. May 24, 2007. Archived on 2024-07-09.
[145]. "Florida gets hydrogen-fueled buses". Drive. 23 May 2007. Archived on 2024-07-09.
[146]. Himmelstein, S. (2023). "Liquid hydrogen system is compact and mobile". GlobalSpec. Archived on 2024-07-09.
[147]. "GENH2 Partners with H2 GENESIS to Provide Small-Scale Hydrogen Liquefaction". Hydrogen Central. 2023-02-24. Archived on 2024-07-09.
[148]. Kurzner, Jeff; Cabus, Nikki (2024-02-28). "FPL announces completion of Florida's ever clean hydrogen hub its kind". South Florida Tech Hub. Archived on 2024-07-09.
[149]. "Stuart Island Energy Initiative". siei.org. Archived on 2013-06-18. Retrieved 2008-05-09.
[150]. "Hydrogen production project promotes green energy transition Việt Nam". vietnamnews.vn. Retrieved 2024-08-14.
[151]. https://news.finclear.tradecentre.io/asx/document/20240722/02830030.pdf.

Chapter (6)
Conclusions

According to Energy Transition Outlook, by 2040, the world's energy system will be distinctly more non-fossil–roughly a 40/60 non-fossil/fossil ratio compared with today's 20/80 ratio. Solar PV will have led the charge for renewables penetration, followed by wind. By then, there will also be a large share of dedicated renewables (almost 7 PWh) for green hydrogen. Fossil sources will still be dominant in the energy mix, with gas exhibiting staying power due to its dispatchability and relatively low emissions, but starting to diminish; oil, one the other hand, will be in sharp decline. This sets up two very important considerations:

- Start planning for less oil. The world has never seen a permanent reduction in oil demand; shrinking demand is likely to gravitate production to lowest cost sources, and a reduction in prices.
- Start planning for New Power Systems. This requires action on grids – in the short-term exploring non-wire transmission enhancement, with longer-term plans for massive grid expansion, and for the introduction of new power market designs to optimize conditions for a high penetration of renewables in the power mix.

The dark horse underlying all of this is AI. It is already clear that AI is playing and can play an increasing role in each step of the energy value chain – including optimized design of renewable component; better renewables siting; enhanced operation of renewable generation; transmission decongestion (for example through dynamic line ratings); demand response; and, not least, end-use efficiency. The potential application of AI is so deep and wide that the world has yet to place any precision on the dimension of its impact. Nevertheless, it will be considerable, if the numbers starting to emerge about the isolated impact of AI on GDP are any indication. Intriguingly, the economic uplift promised by AI is likely to be delivered with low carbon intensity, adding momentum to an energy transition already characterized by improving carbon intensity numbers everywhere.

Ultimately however, it is policy which will have the main impact of decarbonization. One thing we can predict with confidence is that by 2040, the urgency to respond to global warming will have moved far beyond partisan debate. The DNV energy forecast to 2050, that I have used as reference throughout this article, is a 'most likely' forecast of our energy future. It is very far from a 'net zero' future, and hence leads to a global warming of 2.2°C by the end of this century. Limiting global warming below 1.5°C requires a much faster transition. That is still, barely, within the grasp of policymakers worldwide. Should they choose to do so; the trends I have outlined in this brief overview of the energy landscape in 2040 will accelerate greatly. The precautionary principle alone dictates that all stakeholders should start planning for a faster transition.

More Books!

yes

I want morebooks!

Buy your books fast and straightforward online - at one of world's fastest growing online book stores! Environmentally sound due to Print-on-Demand technologies.

Buy your books online at
www.morebooks.shop

Kaufen Sie Ihre Bücher schnell und unkompliziert online – auf einer der am schnellsten wachsenden Buchhandelsplattformen weltweit! Dank Print-On-Demand umwelt- und ressourcenschonend produzi ert.

Bücher schneller online kaufen
www.morebooks.shop

info@omniscriptum.com
www.omniscriptum.com

OMNIScriptum

20 096 142 32

WITHDRAWN